MANAGING TECHNOLOGY DEPENDENT OPERATIONS:
An Executive's Toolbox

Edited by:
Donatas Tijunelis
and
Keith E. McKee

MANAGING TECHNOLOGY DEPENDENT OPERATIONS:
An Executive's Toolbox

Successful use of technology has become the primary factor separating winners from losers in both manufacturing and service industries. Even the smallest of companies has become increasingly dependent upon technology. Technology is everyone's business and cannot simply be left to the technical experts within the company.

The book is intended for non-technical, general management as an overview of practical operations management issues with suggestions of how to deal with the related technologies. It can be looked upon as an executive's toolbox of handy instruments whose application should be easy to recognize.

It has been prepared by practitioners with a total of over 400 years of experience among them directly related to the subject of the book. Each chapter reflects practical experience. The editors, from Lake Forest Graduate School of Management and Illinois Institute of Technology, use their business and technical backgrounds to insert the "glue" that holds the chapters together. The book has been organized to make it possible for the reader to skip around and find the most appropriate tools for their situation.

Editors:
Donatas Tijunelis
and Keith E. McKee

MANAGING TECHNOLOGY DEPENDENT OPERATIONS:
An Executive's Toolbox

Edited by:
Donatas Tijunelis
and
Keith E. McKee

CHICAGO, 2004

Published by

Manufacturing Productivity Center, Ltd.
Suite 504
3115 S. Michigan Avenue
Chicago, Illinois 60616

Copyright © 2004 by Manufacturing Productivity Center, Ltd.

All rights reserved. No part of this work may be reproduced or transmitted in any form or by any means without express written consent of the publisher.

Library of Congress Cataloging-in-Publication Data

Tijunelis, Donatas
  Managing Technology Dependent Operations: An Executive Toolbox/ Donatas Tijunelis and Keith E. McKee
  418 pages
  ISBN: 0-9746354-0-5
    1. Technical management, 2. Change management, 3.New technology, 4 Process and product introduction

& Cover Designed by: Zita Sodeika
        Printed in the Unites State of America
        Richard Printing and Graphics Studio
        8812 103rd Street
        Palos Hills, IL 60465

TABLE OF CONTENTS

PREFACE	EVERY ORGANIZATION FACES TECHNICAL CHALLENGES	7
CHAPTER 1	THE MANAGMENT CHALLENGE - Keith E. Mckee	13
CHAPTER 2	OPERATIONS AND ORGANIZATION – LARGE AND SMALL - Keith E. Mckee	55
CHAPTER 3	WHAT CAN BE EXPECTED OF OPERATIONS MANAGEMENT? - Donatas Tijunelis	71
CHAPTER 4	LEADERSHIP, MOTIVATION AND TEAMWORK - Matthew Puz	95
CHAPTER 5	INNOVATION MANAGMENT - John M. Fildes	153
CHAPTER 6	COMMUNICATIONS – MAKING EVERYONE UNDERSTAND - Michal Safar	187
CHAPTER 7	RECRUITING THE RIGHT PEOPLE - Tim Ryan	233
CHAPTER 8	MANAGING DIVERSITY – TECHNICAL AND CULTURAL - Donatas Tijunelis	251
CHAPTER 9	STRATEGIC ISSUES IN MANAGING TECHNOLOGY - Donatas Tijunelis	281
CHAPTER 10	TECHNICAL SUPPORT TO MULTIPLE OPERATIONS - James P. Nelson	325
CHAPTER 11	PROJECT CONTROL - Eric A. Spanitz	349
CHAPTER 12	LOGISTICS AND THE INTEGRATED ENTERPRISE - Richard Hammond	385
EPILOGUE	SO WHAT DOES THE FUTURE HOLD?	405

PREFACE

If you are looking for practical tools to manage technology dependent operations in a changing economic environment, this book is for you. It is intended to apply to the management of manufacturing, technical services, R&D, information services, engineering, and essentially all organizational functions. It recognizes that whoever is responsible for managing technology is likely to be competing at a frantic pace, pressed for time and constantly searching for new resources. To be prepared is a daunting task. It is easy to be overwhelmed. Under those circumstances, a handy and simple tool is better than one that comes with many options, but its use requires the reading of an instruction booklet with disclaimers and footnotes, all in small print. With this in mind, we focus on the practical subjects and cover them in considerable breadth, leaving the in-depth pursuit of each subject to a list of references and an Appendix. The book is an overview of the field of operations management and, at the same time, a toolbox of handy instruments whose application should be easy to recognize.

By whom and how might the book be best utilized? It is our belief that there are at least two tiers of readers who will benefit from this book. The book is as much for the corporate executive as it is for the functional manager. The one may want to use it as an overview of strategy implementation issues that his or her staff or a subsidiary has to address. The other will pick up tips to more effectively plan, prioritize, motivate, staff, lead and communicate projects in their direct responsibility.

For the executive removed from the day-to-day technical activities by time, location, and perhaps training, there is a need for an overview. This book serves as an introduction to the essentials of strategy implementation throughout technology dependent operations. It is a window to see what it takes to manage technology in practice. It will lead to better understanding of the unique management features of technological programs. It deals with the nagging question of what and to whom to

MANAGING TECHNOLOGY DEPENDENT OPERATIONS

delegate, as well as what to expect. How to motivate toward disciplined innovation without stifling the entrepreneur?

Consider a scenario of a busy executive entering a new position under pressure to be efficient and effective in a technologically competitive environment. The executive may be technically trained, but not practicing, or of a general management background. His or her direct or indirect staff responsible for implementing technical programs consists of several layers of organization, from a dozen to hundreds of people and a similar number of projects. Some of these projects are combined into major thrust programs, some are intended essentially to simply provide maintenance to sustain needed operations, others are yet to be fully initiated. While the fine points of the technology involved may remain beyond the executive's training, the related project management practices should not. If presented in a practical format, even the busiest general manager should easily catch on. Scanning the pages of the book an executive in the position described will quickly recognize the unique features of technology management. He or she will find practical advice on how to structure, prioritize, staff a project, and to whom to delegate its leadership with a reasonable understanding of how technology management will take place on his or her orders.

A functional manager may seek an update of effective practices to support a continuous improvement program within his or her organization. Here the broad range of tips for improving management of technology in an operations environment will be easily recognized by those directly responsible, the project leaders. It can be just what is needed to help explain the process of project justification and economic impact considerations. Often project staff do not recognize project priorities, their relationship to a strategic plan, or at the other extreme, the need for quick and cost effective response to a crisis. These are addressed in the book, as are the needs for creativity and the need to be resourceful.

Consider another scenario: An operations manager has to implement a process improvement for the sake of productivity, while at the same time accommodate on the same equipment relatively crude experiments for an innovative new product that is coming out of a research lab at its infancy. Who has the priority and why? How does the manager explain and motivate the productivity team to accept disruptions

PREFACE

and avoid discouraging the innovator? Will a formalized project schedule help? It is a balancing act for which the book provides a number of tools.

There are many books on strategic planning, technology and project management. Many of these are academic and get into extensive detail. As such, they are useful only to graduate students or strategic planners whose attention is on the fine points of analysis and classification of outcomes. A few are written as briefings from the executive point of view of a portfolio of business units. With respect to implementation, the academic texts extend only up to financial impact issues, while the executive briefings stop at general philosophies without specifics of how things can get done. Furthermore, technology dependent operations depend on project management, but more than that - on project leadership. Unfortunately leadership texts ignore project management tools, while project management texts are written with the technician in mind, i.e. the mechanistic scheduling and cost control. In effect, there is a gap in the tools for managing technology dependent operations. It is a gap in which can be a set of practical tools – a toolbox of practical experience-based management tips.

The contributors to the book are experienced industry managers as well as management-training professionals. They all have learned from theory and, most of all from their practice, the elements of management that work and don't work in technology dependent operations. From the contributors to the book there are reflections of practices at Union Carbide Corporation (UCC), Synergest, Inc., Illinois Tool Works (ITW), AMOCO, Maytag Corporation, Earnst & Young, MPR Corporation, Borg Warner Corporation, US Gypsum (USG), Viskase Corporation, Illinois Institute of Technology Research Institute (IITRI), Dudek & Bock, Wilson Sorting Goods, Continental Can Company, Packer Technologies International (PTI), and DKT Engineering Ltd. The most direct academic input reflects Masters in Business Administration (MBA) and Masters in Manufacturing Technology Management (MMTM) graduate programs and graduate student discussions at Illinois Institute of Technology (IIT), Lake Forest Graduate School of Management (LFGSM), Indiana University, DePaul University, National-Louis University, and Keller Graduate School of Management (KGSM). The Manufacturing Productivity Center at IITRI and their publication of the *Manufacturing Productivity Frontiers* was another source of input.

MANAGING TECHNOLOGY DEPENDENT OPERATIONS

The individual contributions to chapters are as follows: Introductory topics are discussed in the first section of the book. In Chapter 1, Keith E. McKee brings out technology management trends and, as such, introduces the book as a whole. Here you will find discussion on the concept of High Performance Organization (HPO) along with hints for self-assessment with respect to technology management. Technology has to be introduced at a rate faster than can be handled by traditional management approaches. The option is the transformation of organizations into High Performance Organizations where workers at all levels within the company become involved in evaluating, implementing and nurturing new technology. Keith E. McKee, in Chapter 2, deals with the practices of operations and organizations – large and small. There are a variety of organizational types that operate in a variety of ways. The definition of operations and operations management and the responsibilities are varied. Those that are appropriate for large enterprises normally are not suitable for smaller ones. So, what is the scope of operations management? In Chapter 3, Donatas Tijunelis provides an overview of the operations topics that are typically taught in business management (MBA) programs. Here, the readers get exposed in brief to what they would face if they took an operations management course.

After the above introductions, in Chapter 4 Matthew Puz covers the leadership, motivation and team-building basics. Getting people to work together toward common objectives is critical for managing modern organizations. Techniques to achieve this can vary widely and have many names, but the approach is common and can be learned. Here you will find practical advice of the organizational culture required for effective project leadership and team member motivation. Matthew Puz focuses on the primary challenge to get the maximum performance from the available people. This requires the combination of leadership style and appropriate motivation technique. Chapter 5 deals with technical creativity. Getting people to "operate outside of the box" is a major change in all technically based organizations. Here John M. Fildes describes techniques for motivating creativity. Tools are given to help management implement an effective process of innovation.

In Chapter 6, Michal Safar, a technical writer by profession, focuses attention on communications. Communication upward, downward and sideways is critical to

PREFACE

the success of every organization. Added complexity occurs because there is also a need to make all of the information understandable to people at all levels within the organization. Communications become particularly challenging when technical and non-technical people attempt to communicate. Here are some tips for getting your message across. You will find in this chapter a number of helpful suggestions with recommendations for follow-up training. Chapter 7 considers staffing issues from a consultant's point of view. As presented by Timothy Ryan, it should be of particular interest to general management and personnel managers who have to deal with staffing a technical organization. A formal approach is presented for selecting new people who will best fit within the organization as a whole, as well as provides guidance for evaluating people that are being added to a particular group. Donatas Tijunelis, in Chapter 8, addresses the complexity introduced by the diversity of the technical staff with regard to education, experience, and cultural background.

The basic elements of business strategy in relation to technology management are addressed in Chapter 9. Rationalizing the technology advances within the overall strategic plan is critical. Technology can not simply be an add-on to the overall operation of the business. There must be clearly defined operational procedures between the technical managers and the balance of the organization. Here you will find expanded description of projects as part of a strategic program along with tips for project prioritization and means to estimate probable economic impact. Chapter 10 prepared by James P. Nelson extends to the practical issues in managing a technical support function. An approach based on the operation of the technical support group of a large corporation with many operations worldwide is presented. A checklist to help determine which project to do along with techniques for assuring the customer's involvement is presented.

Eric A. Spanitz, in Chapter 11, puts forth the basic elements of "classical" project management and control. These cover the fundamental mechanics of project management, such as work breakdown structure (WBS), critical path scheduling (CPM), progress tracking, and cost control. References are provided for more in-depth follow-up. The information on available and accepted tools for project management is summarized in the Appendix. They should be useful to any practicing project manager.

MANAGING TECHNOLOGY
DEPENDENT OPERATIONS

In Chapter 12, Mr. Richard Hammond describes in effect the current toolbox itself, the very current operations-technology management issues of enterprise resource planning (ERP), system integration, and logistics. He discusses how to deal with the rapidly evolving information technology (IT) — the framework on which modern operations strive for efficiency, quality, innovation, and customer responsiveness.

The editors of *MANAGING TECHNOLOGY DEPENDENT OPERATIONS: An Executive's Toolbox* express their hope that the book is of significant support in your endeavors and thereby will contribute to a more effective use of our combined resources. A shrinking globe cannot afford waste! We will be very pleased and grateful to receive your comments and critiques for future editions.

Editors

Chapter 1 Lead-in:

THE MANAGEMENT CHALLENGE

Successful use of technology has become the primary factor separating winners from losers in both manufacturing and service industries. Even the smallest of companies have become increasingly dependent upon technology. Technology is everyone's business and cannot simply be left to the technical experts within the company.

The following chapter discusses general trends in technology management. It points out what is happening to management challenges as a result of technological advances. It doing so, it introduces the book as a "toolbox" to provide managers with an overview of the practical basics and the challenges involved.

Organizational and personal self-assessment are covered in this chapter. It raises the point that in the recent past mangers were interchangeable. That, except for those responsible for very specialized areas, "A competent manager can manage anything." That is only true if the manager can appreciate the scope, level of effort, and the impact of what needs to be done and to whom the work should be assigned. Most of all, the manager must relate across organizational boundaries and professional disciplines. The chapter addresses the management of change and leads into the management issues associated with rapid acceptance of the "high performance organizations" (HPO). This then transitions into Chapter 2 where the corporate organizational structure is discussed in greater detail as it relates to operations management.

MANAGING TECHNOLOGY
DEPENDENT OPERATIONS

CHAPTER I

THE MANAGEMENT CHALLENGE

Keith McKee

Management of technology has become the primary challenge for managers at all levels in every sector of the economy. New technology results in changes impacting equipment, organization, and people – many long established management approaches are no longer applicable. Management of change has become the major challenging facing companies worldwide.

Not too many years ago, the management of technology was primarily the purview of the R, D&E Directors with top management involved primarily in controlling the budgets allotted for R, D&E. Management of R,D&E which has to be directed by a technical specialist was accepted as different than management of traditional operations. An extensive body of literature specifically for R, D&E management developed on the assumption that these operations were different from conventional management. This model assumed that management within the balance of the organization could then simply deal with the R, D&E specialists. This model is no longer applicable. The penetration of new technology into every aspect of the corporation, along with the very rapid rate of technology change, means that every manager is a "manger of technology dependent operations."

THE MANAGEMENT CHALLANGE
By Keith McKee

Traditional management approaches, which remain the basis for much of today's education, are based on an earlier age when technology change was relatively slow. In the "earlier times," it was reasonable to expect that an operation would remain unchanged over many years. Procedures could be documented and optimized, standards established, and workers trained to do specific functions. The "mangers" knew

> ....*every manager is a "manger of technology dependent operations".*

what should be done in detail and were able to direct the workforce in detail. A manual of Standard Operation Procedures (SOPs) effectively made every decision in advance so that employees were simply human machines preprogrammed to perform certain functions. Traditional management was organized like a military organization with orders coming down and information flowing upward along formally established lines. Except for those managers responsible for specialized areas, it was assume that mangers were interchangeable. "A competent manager can manage anything."

The rapidity of technological change has lead many organizations to experiment with various new forms for their organizations. Such alternate organizations come in a variety of forms and with various names – High Performance Organization (HPO), Participative Management, World Class Organizations, etc. For the purposes of this chapter, HPO is used with the understanding that this is a generic term for this new type of organization. The driver for these changes has been the rapid evolution of technology, which requires a more agile organization and active involvement of employees at every level. "Hierarchical" organizations lack the flexibility to handle rapid technological change.

The common challenge for all of these evolving organizational structures is "Managing Technology Dependent Organizations." As mentioned earlier, traditional management was typically very

> ....*challenged by the dichotomy between two management systems.*

different from R, D&E management. Managers within the newly evolving organizations are being challenged by the dichotomy between these two management systems. It is the authors' belief that the first requirement to integrate these two management systems is an understanding of both. For that reason the "Toolbox for Managing Technology Dependent Operations" provides managers with an overall understanding of the challenges involved.

Each reader comes to this book with different experiences and education. In recognition of this, the authors and editors have tried to make this book readable in any order and/or as separate articles. The analogy to "TOOLBOX" is appropriate – readers can select the tools that they need without unloading the entire toolbox. The balance of this chapter introduces some of the trends and issues in "Managing Technology Dependent Operations" for all levels throughout the organization.

TECHNOLOGY

Everyone has his or her personal understanding and definition of technology and of the role that technology plays in society. The rate of technological changes has accelerated over the past decades and there is every indication this pace will continue or more likely increase. The public is busy discussing and reading about the "next big" technical thing - cloning, nanotechnology, designer materials, space, artificial intelligence, etc.

New technology is very difficult to define...

New technology is very difficult to define and definitions can be very elusive. What is considered "very advance" today may well become commonplace next year. All of the tools, that we presently have easily available as package software attached to icons on every personal computer, were being studied as advanced research topics less than a generation ago.

THE MANAGEMENT CHALLANGE
By Keith McKee

In technical circles AI stands for "artificial intelligence" but not all that humorously it has been suggested that it could equally stands for "anything interesting." A wide variety of research ideas suddenly became AI when AI became a "hot" research subject. These phenomena applied to all advanced technologies – as a new technology starts, to get attention in the press, both the public and researchers rapidly add to that new technology subjects of interest from the past, present, and future.

Technology changes are fairly obviously and well known within our personal lives. A large portion of the technology, used in our everyday lives, was not available only a generation or two ago. Much less discussed, but equally or even more impressive, are the technologies that have entered the commercial world. Since it impacts individuals directly, the changes in the financial service industry are very obvious. Many people no longer ever enter a "brick and mortar" bank for any purpose. People still use banking service, but their funds are directly deposited, bills are paid on the computer, money is obtained from ATMs, etc. The extent to which the individual's interactions with financial institutions have changed is really the "tip of the iceberg" – the industry itself has been almost totally restructured because of all of this new technology. Financial institution now employ very sophisticated technical staffs to develop and maintain their systems. Everyone reading this book has learned to interact with banks in new and different ways that have been driven by technology change.

Even more hidden from the public are the changes that have happened within almost every industry. As one example oil refineries have effectively become large computer controlled operations with simulations monitoring and controlling every aspect of the operation – humans watch what is happening but only get involved when the automation fails. Perhaps it sounds like science fiction to suggest that automation will soon be able to maintain itself,

> *...automation will soon be able to maintain itself...*

MANAGING TECHNOLOGY DEPENDENT OPERATIONS

but it should be kept in mind that is what is happening today was considered "far out" only a few years ago. Perhaps the refinery a generation in the future will have a room filled with robots playing cards and drinking coffee that are available to be dispatched as required to maintain and even upgrade the refinery.

Technology has had a dramatic impact on all types of business. People are actually better able to deal with new technology then are companies. An individual or a family can in most instances simply not use new technology – they can deal with their bank in person, not use a computer, not have a cell phone, etc. By choice, individuals and families can delay getting involved with a new technology until they have become sufficiently familiar with it to be comfortable. Organizations that take these same "not for us" or "wait and see" attitudes may well cease to exist. Pressure from customers and competitors simply will not allow an organization to ignore new technology. For competitive reasons, organizations are frequently required to become involved with new technologies before either the organization or the technology is ready.

> *...new technology may not do what is hoped and it frequently doesn't.*

Individuals and families have another advantage – by the time a new technology is available to them it is usually fairly well packaged and reasonably user - friendly. Organizations normally have to move into new technologies when they are much less well defined. An obvious example here is that an organization that wants to develop a new technically oriented product or service is very likely to get involved in actually developing that technology. One can easily imagine the technology challenges faced by a financial institution before they could offer web-based baking. There are also major risks that the new technology may not do what is hoped and it frequently doesn't. A current example, where the final results are still unclear, is high definition TV. Large

THE MANAGEMENT CHALLANGE
By Keith McKee

expenditures of time and money have been made, but thus far the market has not followed.

What is technology? As used in the title of the book, we would argue that "technology is the application of any new knowledge to real world problems." The International Technology Education Association (ITEA) defined technology as "changing the natural world to satisfy our needs." These two definitions are sufficiently similar for our purposes. The implication of both definitions is that changes are being made to alter the present status.

Many people think of technology as simply another term for science or engineering. A 2002 Gallop Poll (1) for ITEA found that 59% of the respondents said that "science and technology are basically one in the same thing" and 61% said, "engineering and technology are basically one in the same thing." In the same poll, 67% of respondents said that the first thing that comes to their mind when hear the word "technology" is computers. Computers are important, but only part of the new knowledge that is driving change. Similarly we would argue that not all technology needs to be science or engineering based. It seems pointless to get involved in semantic debates on these definitions. For the purposes of this book

TECHNOLOGY IS THE APPPLCATION
OF ANY NEW KNOWLEDGE
THAT LEADS TO CHANGE
WITHIN AN ORGANIZAITON

Basically, if some knowledge or information new to a company leads to change within that company that knowledge is "technology." For the purposes of this book the key point is that new technology requires change.

PROJECTS

The word project is widely used and has many different meanings. "A scheme or plan" is the definition given in New Webster's Dictionary 1998 edition. "Plan" might be the preferred definition because "scheme" has some negative connotations. In practice, we rarely use the word project to simply mean a plan. When someone talks about a project to "put up a stone wall," everyone imagines that that person is going to purchase stone, mix mortar and build the wall. The image that comes to mind is much more than a plan would seem to imply. For business purposes a more useful definition might be that

A PROJECT IS AN ACTIVITY WITH A BEGINNING AND AN END

Again for business purposes, this definition is often modified with phases like

DONE WITHIN LIMITED TIME AND FUNDING

Figure 1: The Project

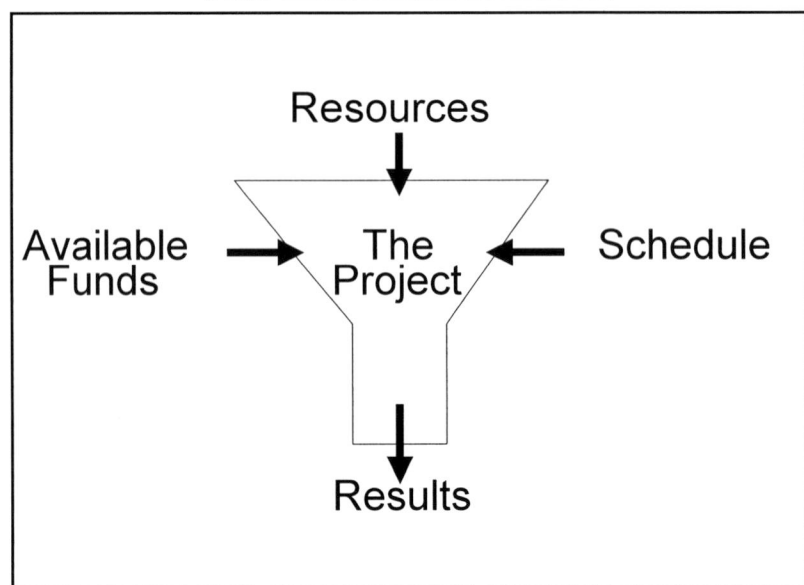

THE MANAGEMENT CHALLANGE
By Keith McKee

This definition of a project is general enough to be applicable to almost any activity. A project can be to read a given book or to build a new airport. Clearly the complexity of these two projects differ significantly. As might be expected then, these two projects would be approached in totally different ways. The project plan to accomplish the book reading project might be as simple as reading one chapter each night before going to bed. The project planning for the airport would involve multiple activities done by divergence organization over many years or even decades. Building an airport would be a "Super Project" within which there would be numerous major supporting projects, e.g. purchasing the land, designing the airport, constructing the runways, building the control tower, etc. Taken on their own, each of these supporting projects would be very major projects that could take years and under which there would be many support projects.

Within organizations, projects can come in various shapes and sizes depending upon both the business of the business as well as its organization. A single person, working on a part time basis, may serve as the project leader as well as the entire project staff for some projects. Other projects will involve an individual directing numerous other people and/or organizations working on the project.

Over this range of possibilities, the person leading the project hopefully has both the responsibility, as well as the authority, required to achieve the

> *Projects are typically cross-functional activities.*

project goals. Many projects fail, or at least falter, because management does not support the person nominally running the project. Simply telling a person that he or she is a project manager without giving the power required to achieve the goals of the project is frustrating to the project leader as well as the entire organization.

Projects are typically cross-functional activities involving people who do not normally directly report to the project manager. Availability of the project staff depends on the cooperation and active involvement of those to

MANAGING TECHNOLOGY DEPENDENT OPERATIONS

whom that staff directly reports. It is very common that the line management will perceive that the work within the home department is of higher priority than the project work. Under these conditions, people best able to contribute to a project may not be fully available or alternately "second-rate" people will be assigned to the project. If the person assigned to a project is pressured by their normal manager (the one who gives raises and promotions) to emphasize their normal assignment most staff will honor that request no matter how important the project may be to the company.

Another very common problem is resource competition between projects. When multiple projects are underway within an organization, critical individuals or other resources may simply not be available to work on all of the projects as desired by the individual project leaders. Clearly higher management must intervene to resolve these issues. Hopefully management attention comes early enough so that actions can be taken to avoid these problems, rather than coming after the problems have already created crises.

> *Standard Operating Procedure (Bureaucracy) frequently makes the project mangers job very difficult.*

Standard Operating Procedures (Bureaucracy) frequently makes the project mangers job very difficult. SOPs for purchasing and subcontracting may prove to be major challenges for a project leader. Often project managers are not given the signature authority and have to depend on line management cooperation. When the timing of a project is critical, the normally prudent requirements for obtaining multiple bids can be a problem. Similarly, approvals required for overtime or business travel can limit the flexibility available to the project manager. Again these are issues that require the involvement of higher level of management before they become problems.

A PROJECT MANAGER WHO DOES NOT HAVE CONROL OVER REQUIRED RESOURCES IS NOT MANAGING THE PROJECT

THE MANAGEMENT CHALLANGE
By Keith McKee

For an organization with only a few projects underway at any given time, upper management is knowledgeable and involved so the above challenges are typically handled with relative ease. This is how traditional organizations operated. As the number of projects increases, it becomes much more difficult for upper management to solve these problems as they arise. If management does not recognize this challenge and established in advance procedures and controls to handle such situations routinely, there will be "project gridlock." Project gridlock, like traffic gridlock, is much more difficult to untangle than it is to avoid.

In organizations where teams and teamwork are emphasized, there is frequent reference to "team projects." Perhaps there is a way to actually do this, but in this writer's view this is an oxymoron. A team can be very valuable in developing and planning a project, but project execution requires one person as the project leader. For a team, this person could be the team leader, team facilitator, or the manager responsible for the team. When a team project is undertaken this point should be clearly established in the beginning. The old saw of "too many cooks spoiling the stew" would seem very relevant without this clarification.

> *...project execution requires one person as the project leader.*

Projects, that do not have clearly defined objectives to be accomplished within limited time and funding, are not projects as considered here. Management frequently uses the word "project" with many different meanings. For example, a person can be told that as a special project for the boss they should evaluate the benefits of a new technology or evaluate a competitor's approach to particular situations. These assignments may be important to the organization, but to the extent that they are not clearly structured they should not be treated as "projects." As was noted earlier "A Project is an Activity with a Beginning and an End Done within Limited Time and Funding."

**MANAGING TECHNOLOGY
DEPENDENT OPERATIONS**

Because of the importance of projects to operations in implementing technology change, there are further discussions of very specific project management tools through the book.

TRADITIONAL MANAGEMENT

Until relatively recently almost all organizations were based on the military model – a model that some people claim can be traced to the Roman Legions. The organization chart for these "hierarchical" organizations can be shown as a triangle with one person at the apex and with subsequent layers of management imposed between top management and the workers. In this model, each person has one boss and only one boss. The entire organizational chart can be described by documenting the reporting hierarchy. (See Fig. 2)

Under traditional management a manager has almost total responsibility for and control over those individual reporting to him or her. The assumption was that the boss knew more about the activities of that group of individuals then anyone within the group. Those reporting to the "boss" were expected to carryout the leader's orders. Results were reported to the boss and if problems occurred, it was up to the boss to resolve them. Intermediate levels of management could act only within clearly defined bounds. By the time this hierarchical control reached the bottom of the organization, those workers were primarily expected to show up on time and do what they were told. Some people referred to this as "punching the time clock and leaving your brains on a hook at the door."

THE MANAGEMENT CHALLANGE
By Keith McKee

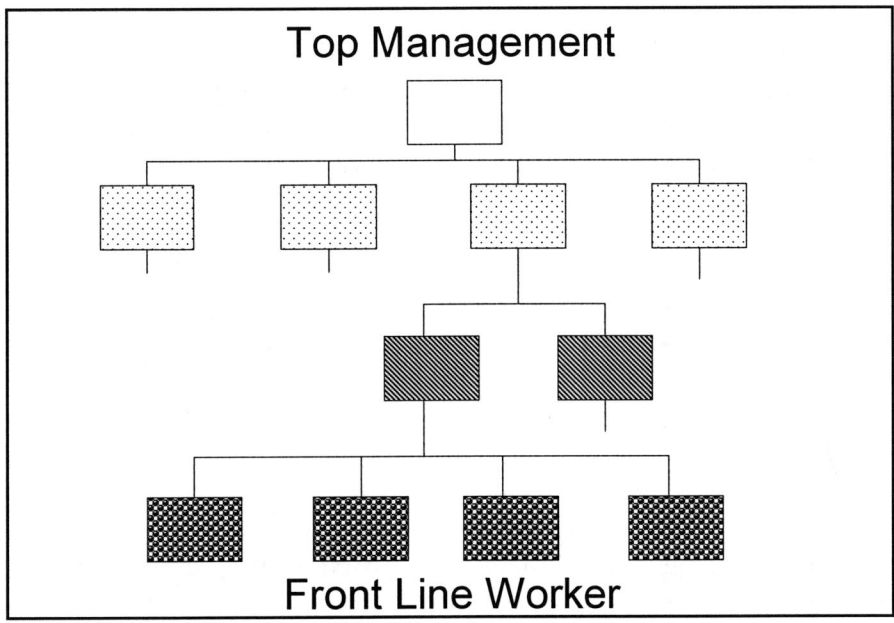

Figure 2: Hierarchic Organizations

Most traditional organizations over period of many decades operated as steady-state operations. Whether producing "widgets" in a manufacturing plant or handling money in a bank, the procedures or transactions remained effectively unchanged over generations.

TRADITIONAL = STEADY STATE

Obviously nothing is every totally steady state – there were at least occasionally some improvements or changes introduced. New equipment, a building addition, a new product, etc. were introduced. Nevertheless, at a given position in time, these changes were

> *...large portion of the existing management literature is based on steady state operations.*

MANAGING TECHNOLOGY DEPENDENT OPERATIONS

barely detectable. Operations during one month were effectively the same as operations a month earlier. Changes that were made were discrete events in the history of the operation and handled as deviations from the ongoing steady-state operations.

When charges did occur within steady-state operations, it was a major deal that received top management attention. Management selected the changes to be made from several candidates generated by the staff. Funds were budgeted for the planned change and a member of the senior staff given clear responsibility for this "project." Since traditional (steady state) management was the norm for several decades, a large portion of the existing management literature is based on steady state operations.

Some professionally oriented activities were always included within traditional organization as service activities. The Legal Department, the Accounting/Finance Department along with Research, Development and Engineering (R, D &E) activities were normally managed by specialists from those professional activities. Within recent years, Information Technology (IT) has emerged as another specialty. Managers responsible for such "service" activities were normally professional in the area that they managed. Typically these managers and their activities were treated as external to the core organization. Budgeting for and control of these service activities was the responsibility of management. The balance of the organization normally accepted them as services that did their assigned functions but were not part of the core business.

> *In-house project based groups have a single customer – the parent company.*

The R, D&E as well as the IT organizations typically operated using a project management approach internal to those organizations. The management style that developed for these service organizations was normally very different from what was used by the organization of which they were a part.

THE MANAGEMENT CHALLANGE
By Keith McKee

This was not surprising since effectively all of activities within these groups were handled as projects.

Management of these in-house service organization developed systems for managing a portfolio of projects to satisfy corporate management. Employees within these groups were either project managers or working on projects. Some of the management tools used by these in-house organizations were taken from the formal project management literature, but much was developed specifically for these service functions. There is an abundance of literature on R,D&E, Product Development, and IT management.

The difference between these in-house project oriented groups and independent organizations that operate using the project management approach are worth noting. Engineering consultants, construction firms, contract R,D&E organizations, and many government agencies also operate with essentially all activities done as projects. There is a significant difference between managing these commercial, project-based organizations and an in-house project based organization.

> *...externally funded organizations have a different situation - each project is for a different customer.*

In-house project based groups have a single customer – the parent company. Management of these in-house service groups, handles a portfolio of projects funded by the same customer – the parent company. Under these circumstances, management has considerable discretion in establishing priorities among the projects within their portfolio.

Managers of commercially funded project organizations have a different situation. Since each project is for a different customer, management

MANAGING TECHNOLOGY DEPENDENT OPERATIONS

has little ability to reschedule or prioritize for the overall benefit of their organizations. Results have been promised to the customer with a schedule and costs established. Any changes that management is forced to make to balance internal workload are failures that result in unhappy customers. Because of these differences, the tools that have been developed and are used for in-house organization tend to differ substantially from those used by commercially oriented project-based organization.

ORGANIZATIONS UNDER PRESSURE

Steady-state (transaction) based companies, as described above, handled new technology as improvement to the steady-state operations. Technology was introduced deliberately with top management maintaining control and with the new technology implemented by managers operating within the existing chain of command. From the point of view of those at the bottom of the organizational pyramid, the new technology comes as changes made by management without worker involvement anywhere within the decision making process. Depending upon the change, the workers (and/or their union) accepted the change or put up a fight. Those at the bottom were simply pawns in the game of business.

> *...many steady-state operations started to "boil over"... Change simply could not be handled from the top down.*

Starting in the 1990, many steady-state operations started to "boil over." New technologies driven by quality, computer technology and international competitiveness, etc., were coming on more rapidly than these traditional organizations could handle with management controlled projects. Change simply could not be handled from the top down as improvement to the steady-state situation. At the same time, companies based on major new technologies such as biotech, communications, and computers were being started. For these start-ups, steady-state was a

THE MANAGEMENT CHALLANGE
By Keith McKee

meaningless concept.

The results have been a very rapid transition toward "high performance organizations." HPOs are organizations where everyone in the organization is empowered and involved in new technology and change. In an HPO, workers at every level within the organization do many activities that would traditionally have been handled by management. HPOs operate under a variety of names including Worldclass, TQM, Six Sigma, Baldrige, and ISO. Details and terminology are different, but the common thread is that the traditional hierarchical management structure is giving way to participative management with decisions being made at every level within the company.

> *...HPOs operate under a variety of names including Worldclass, TQM, Six-Sigma, Baldrige, and ISO.*

This charge was recognized by Marc Tucker, President of the National Center on Education and the Economy early in the 1990 when he said,

"The top businesses are finding that to get very high performance out of an organization, they must assign the frontline workers duties and responsibilities that typically have been assigned only to management and senior professional personnel. Frontline workers need the authority to get the job done and to figure out how to do it in a context in which the goals have been clearly specified and agreed to: we must give them the resources they need to do the job and to hold them accountable for their performance" (2)

Tucker's views were one of the first statements of this major philosophical change. Since then many companies have moved toward becoming HPOs. This transition is still underway, but it seems clear that this is the future direction for leading organizations in all fields. It is important to note that Tucker said that involving front line workers was "necessary not nice."

MANAGING TECHNOLOGY
DEVELOPMENT OPERATIONS

"Necessary not nice" is a very important point and represents a major change. Literally for several decades, social scientists had been talking about improving the "quality of worklife" (QWL) in order to make work more interesting and less boring. In the 1970s, there was a Presidential Commission on the "Quality of Worklife" which was dedicated to moving in this direction(improving the QWL). There was at that time a group of organizations located primarily at Academic Institution that were dedicated to improving QWL. Many people and organizations thought that this was a good idea, but it never became widely accepted. Labor (and specifically unions) felt that somehow this was an attempt by management to get greater cooperation from the workers by "conning" them with kindness. Management was equally uninterested - they saw extra activities and costs without clear benefits. In spite of all of the activities to promote this approach, there was no definitive data to show that QWL positively impacted companies. QWL more of less faded away with some of the centers disappearing while other moved into Productivity and/or Quality Improvement. QWL might have been "nice" but it was not "necessary" and there was nothing to indicate that it improved the organizations in which it was implemented.

HPO is very similar to QWL and uses many of the same techniques. The difference is that HPO was "necessary not nice". The combination of accelerated technological change, emphasis on improving quality, and increased competition both internationally and locally forced organizations to move toward becoming HPOs. HPOs are becoming the de facto organization of choice.

HPOS ARE EMERGING BECAUSE THEY ARE NECESSARY,
NOT BECAUSE THEY ARE NICE!

...HPOs did better than more traditional organizations...

Documentation that HPOs did better than more traditional organizations began to appear in the mid 1990s. The Indiana Labor and Management Council

THE MANAGEMENT CHALLANGE
By Keith McKee

demonstrated that "HPOs were more successful." The following quotations are from a March 1999 Report (3)

"In 1992, working with the Manufacturing Productivity Center at the Illinois Institute of Technology, the ILMC conducted a survey of over 600 companies. This study produced the first clear evidence that <u>social system strategies had greater impact on a company's success than technology/ equipment or technical system interventions.</u> (Underline added by writer for emphasis) This was an important finding that conclusively showed that a company could not ignore the "softer systems," but rather that they are critically important to success. Based on this study the ILMC produced an initial high-performance model that illustrated the "fit" among technology and equipment, technical support systems and social systems with an organization's business goals."

Figure 3: Organizational Fit

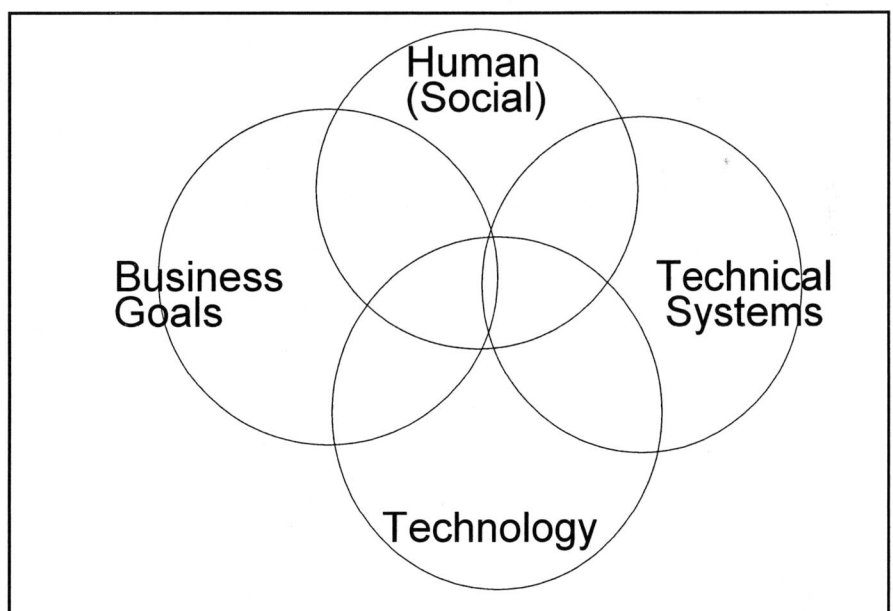

MANAGING TECHNOLOGY
DEPENDENT OPERATIONS

Top management did not change their approaches because of this or similar surveys. The transition to HPOs started because companies could no longer operate under the traditional top-down management structure. The barrage of challenges and changes were simply so numerous that the old model could not handle the internal pressures and something new was required - HPOs, by whatever name, have been the results. Since management could not keep up, they had no choice but to restructure the organization to get more people involved. Startup companies arrived at the same position directly since they were never sufficiently stable to operate in a steady state mode.

In HPOs, people at every level are involved in dealing with customers as well as in making decisions and acting upon those decisions. This is a far cry from the military model. One way of picturing the charge is to consider the organization chart as an upside down triangle with the apex (and top management) at the bottom. In this idealization, the first line workers are "key players" with their managers supporting these workers. Based on this model, first line supervisors become coaches or facilitators working to support rather than direct the front-line workers. This logic is "carried down" to the higher levels of management each of which supports those people who report to them.

Figure 4: Idealized HPO Organization

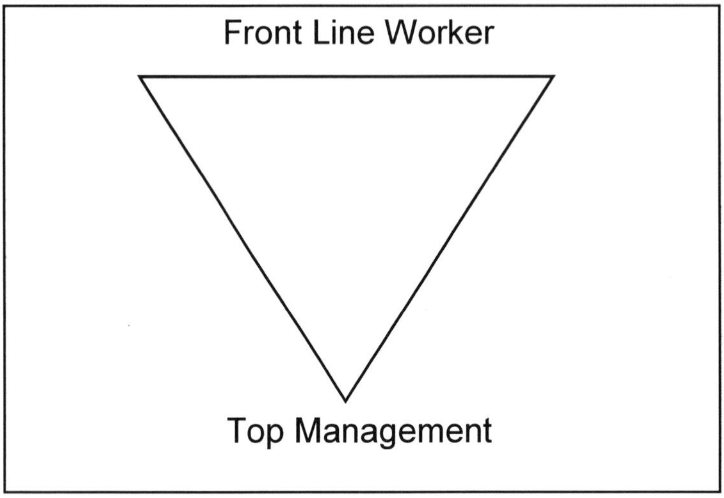

```
┌─────────────────────────────────────┐
│          Front Line Worker          │
│         _____         │
│         \                 /         │
│          \               /          │
│           \             /           │
│            \           /            │
│             \         /             │
│              \       /              │
│               \     /               │
│                \   /                │
│                 \ /                 │
│          Top Management             │
└─────────────────────────────────────┘
```

THE MANAGEMENT CHALLANGE
By Keith McKee

This exaggerated image (and it should be understood this is an exaggeration) of the inverted triangle as the new organizational structure does emphasize the fact that organizations have become much more dependent on workers at every level. As the pace of change has accelerated, organizations have had to depend on decisions and actions by individuals working individually and/or as teams. This is a "major sea change" in what is expected of employees.

These changes are impacting the education and training being developed for the workforce. As one example of this change, the new standards developed for front line workers as well as their supervisors are based almost completely on those skills required within an HPO. At the college level, there are also strong movements to acquaint students with "living within an HPO."

PEOPLE

"People are our most important asset" is frequently said, but normally not believed by the management or by the people themselves. When the first action taken by a company having problems is slashing the workforce, this is a clear demonstration that this is an empty phases. There are a few companies that very seriously treat people as their most important asset. As an example, Lincoln Electric, a billion dollar a year company with 7000 employees founded in 1895, claims to have not laid off a worker for any reason other than cause. Many companies that made similar claims only a decade ago have had to eat their words with massive bloodletting.

> *...few companies seriously treat people as their most important asset.*

With this observation in mind, HPO management of technology based operations has a very interesting challenge. People absolutely positively must be involved if the organization is going to be able to adapt to rapid technical

MANAGING TECHNOLOGY
DEPENDENT OPERATIONS

> *People* **absolutely** **positively** *must be involved...*

change. At the same time, the managers are not in a position to assure the people (including themselves) that they have job security. This is a very conflicted situation - the organization needs people to commit all of their effort to benefit the organization and yet organizations are not in the position to guarantee that these people will have jobs in the future. As strange as it sounds, this is the type of social contract that is being required as organizations become HPOs.

This may sound like *Alice in Wonderland* but perhaps it is not as crazy as it first appears. Everyone recognize this reality for start-up technical companies. People go to work for them, give 120% effort and hope that when and if the company succeeds they will become rich. The potential of significant fiscal rewards is important, but it is not the only motivation for people to join such organizations. Even if the company failed and there was no "pot of gold," many of these people gain experience that would have taken many years in a different environment. Many of them also simply had "fun."

The "climate" in most of these organizations is very positive. Climate as used here is based on the definition developed by Goran Ekvall of the Swedish Council for Management and Organizational Behavior (4).

Nine Climate Dimensions

- **Challenge and Involvement** - *represents the degree to which people are involved in the daily operations, long-term goals and visions of the organization*
- **Freedom** - *where people have independence and autonomy to define much of their own work and exercise discretion in their day-to-day work activities*
- **Trust and Openness** - *refers to emotional safety in relationships without fear of reprisals and ridicule in case of failure and where communications is open and straightforward*

THE MANAGEMENT CHALLENGE
By Keith McKee

- ***Idea Time*** - the amount of time people can use (and do use) for elaborating or developing new ideas
- ***Playfulness and Humor*** - the spontaneity and ease developed in the climate
- ***Conflict*** - refers to the presence of non-productivity personal, interpersonal or emotional tensions in the workplace (Writer's note - this is only one of 9 factors which is considered as negative)
- ***Idea Support*** - involves the ways new ideas are treated
- ***Debates*** - involves positive encounters, exchanges of viewpoints, ideas, and differing experiences and knowledge, and useful challenge to each other's thinking
- ***Risk-taking*** - refers to a tolerance of uncertainty and ambiguity exposed in the workplace where people feel that they can "take a gamble" on some of their ideas

It is easy to see why people might enjoy working in an organization with a good climate. Many start-up technology companies have a very positive climate. A reasonable question is whether a good climate can work in other organizations. The Indiana Labor and Management Council survey conducted in 1998 with the Chicago Manufacturing Center considered this question. Robert J. Firenze in *Moving Toward a High Performance Organization* (3) summarized the result of the survey by saying,

"Based on the reported data from company presidents and general managers it appeared that there was a significant relationship between working climate and product/service quality, profitability, productivity and degree of market share."

MANAGING TECHNOLOGY
DEPENDENT OPERATIONS

A GOOD CLIMATE WITHIN AN ORGANIZATION MAKES THE PEOPLE HAPPY AND AT THE SAME TIME IMPROVES THE ORGANIZATION'S PERFORMANCE

A good climate correlates with positive business results. Perhaps a new social contact is possible - a contract that says that if an organization develops a positive climate that the people will help the company to succeed. The question really is not whether this is possible but rather how it is being achieved.

Some of the major changes which are underway are the redefinition of workplace skills. No longer are the primary skills that are valued only those related to routine manual or clerical activities. Rather, the skills that are being emphasized and developed are those required to operate in an HPO. The National Skill Standards Board (NSSB) is developing skill standards for various trades. For example, the Manufacturing Skill Standards Council (MSSC) is developing the skills required for working in manufacturing. The critical work functions for this standard are:

1. *Produce Product to Meet Customer Needs*
2. *Maintain Equipment Tools and Workstations*
3. *Maintain a Safe and Productive Work Area*
4. *Maintain Quality and Implement Continuous Improvement Processes*
5. *Communicate with Co-Workers and/or External Customers to Ensure Production Meets Business Requirements*
6. *Coordinate Work Team to Produce Product*
7. *Ensure Safe Use of Equipment in the Workplace*
8. *Correct the Product and Process to Meet Quality Standards*

These standards are being developed for entry-level workers as well as front-line supervisors. These standards, which target high school and community college students, are a far cry from the previous standards which

THE MANAGEMENT CHALLANGE
By Keith McKee

were based almost totally upon training people to operate specific equipment or perform specific functions with almost no attention based to overall operation, communications, etc. These standards are based much more on basic academic skills rather than on clock-time training. Further, they emphasize interacting with others rather than standing at a machine or exercising a very specific skill.

> ...*every (IIT) student must participate in two IPRO projects... real-world challenges provided by local companies..*

Similar changes are occurring at the college level. In a variety of ways, schools are working to breakdown the "walls" between departments, schools, etc. Until a few years ago, it was possible for a student to complete a degree in business, engineering, or law without ever meeting a student or professor from any other discipline. This is not the real world that these specialists are likely to find when they go to work. Various approaches are being implemented within universities to break down the "walls."

As an example, the Illinois Institute of Technology (IIT) in Chicago has developed a program that requires that every undergraduate before graduation participate in two InterPROfessional projects. IPRO projects are typically real-world challenges provided by local companies. Teams of 10 to 15 students from various undergraduate and graduate specialties (and from various academic levels) work together to solve this real world problem. It is up to the team to define the problem, pursue the solution, and present the results. Each semester, the first few weeks are near panic. Students who have not participated in an earlier IPRO are waiting for someone to give them an assignment within their area of specialty.

There are also major communication problems. With panic as a motivator and with a faculty mentor asking questions about schedule, about 1/3 of the way through the semester things start to come together. The results of most of the IPRO end up with very impressive results. Of even greater importance, the students suddenly understand what is involved in and how to work together. The results are students who are prepared to move into the real world where teams, communication, and interdisciplinary activities are the norm.

> *Organizations are retraining people already within the organization.*

Training standards and college curricula reflect the changing role of people within HP organizations. These changes are being driven by what real organizations require. These changes within education and training are still cutting edge so the number of people prepared via these routes is very limited. Organizations cannot wait to hire people that are prepared for this paradigm, but rather must retrain the people already within the organization.

Organizations that are moving toward becoming HPOs, via TQM, Six-Sigma, Baldrige, ISO or many other paths; are retraining their workforce and their managers for this new management paradigm. In the best of these organizations, both management and the workers have changed tremendously. At least in part based on this new approach, the best of these organizations have been able to weather rough economic situations without losing the support of their people.

ORGANIZATIONS

In earlier sections, a variety of organizations have been mentioned and discussed. In the following table five organizational types are profiled. For the purposes of this presentation, "in-house" service organizations are different from the organizations within which they operate since they normally operate with completely different management styles. "In-house" and "contract" service organizations are also considered separately since they operate with different measures of success and hence different management approach. The comments under each organization provide indications of the differences and similarities.

THE MANAGEMENT CHALLANGE
By Keith McKee

ORGANIZATIONAL TYPES

- ❏ Traditional - Steady-state operations - service, government and manufacturing
  - Hierarchical management
  - Informal project management
  - Slow changing
  - Changes controlled by management
  - Low worker involvement
  - Few teams
- ➢ *Limited number of project directed by management*

- ❏ Project Management - Government and construction
  - Hierarchical management
  - Formal project management tools and techniques
  - Fast changing
  - Professionally oriented
  - Low worker involvement
  - Teams at professional level only
- ➢ *Every thing handled as projects with formal tools*

- ❏ Contract Service Organizations - Commercial & government laboratories, consulting organizations, professional organizations
  - Technical Management
  - Formal and informal project management tools and techniques
  - Funded project for external customers - time and cost constraints
  - Professional Oriented
  - Involvement of all professional staff
  - Teams involving professional staff
- ➢ *Everything is a project using various tools depending upon customer and project manager*

- ❏ In-house Service Organizations - Corporate operations for research, new products, IT, manufacturing, etc

MANAGING TECHNOLOGY
DEPENDENT OPERATIONS

- Technical Management
- Formal and informal project management tools and techniques
- Internally funded project - interplay and tradeoffs between
- Technical and Corporate management
- Professional Oriented
- Involvement of all professional staff
- Teams involving professional staff

➢ *Everything is part of the corporate portfolio of projects - individual companies have their systems for portfolio management. Formal and informal project management tools*

❑ High Performance Organizations (HPOs)
- Participative management - everyone involved
- Formal, informal, and casual project management tool and techniques
- Competing projects generated throughout the organization
- Everyone involved
- Teams involving everyone

➢ *Projects coming from management, customers, and teams compete for attention and resources*

It seems unlikely that any organization operates solely within one of these organizational types. Rather "real world" organizations, operating within one of these organizational types, will have overtones of other types. In some instances an organization will be a hybrid of two or more organizational types. Hopefully, this table will help readers to evaluate their current organization as well potential directions for change.

MANAGERS

Managers are those men and women who manage operations. Managers of specialized operations are typically technical specialists in the areas that they are managing. Earlier, managment of internal service activities was specifically considered. Managers of RD&E, Product Development, Accounting, IT, and Legal Department normally fall into this category.

THE MANAGEMENT CHALLANGE
By Keith McKee

Management of these and other specialties can be effectively done only by an expert in the specialty involved.

> *Utility managers are able to move into a variety of slots...*

Utility management is used here to refer to those men and women who are managing operations at all levels. The managers within the core activities of traditional organizations are utility managers. They also fall in the category of General Management. These are the managers (and for the balance of this section, they will simply be referred to as managers) who typically can be reassigned to another activity within the organization with limited difficulty. Those people managing your local gas, electric, and water companies may be utility managers, but not because they are managers of utility businesses. The analogy being used here is to a "utility" baseball player– a player that can handle a number of positions. Utility managers are able to move into a variety of slots within their organization.

Very frequently the key to the success of these managers is their knowledge of the organization or industry in which they operate. The education, training, and experience of these managers vary widely. Their understanding of and approach to new technology covers the full range of possibilities. Similarly their knowledge of and experience working with different types of participative management systems will include every possibility - some will be comfortable having all of the workers involved while others will still be hierarchical managers.

Table 1 provides a tool for profiling a manager. The reader can use this scoring system for themselves, their subordinate, their superiors, and their peers. The scoring system is such that as scores approach 5, the manager is participative, technology literate, and generally operating as a manager of a high performance organization. As the scores approach 1, the manager is operating as a hierarchical manager. The assumption in this book is that 5 is better than 1, but clearly in some organization that assumption might not be appropriate.

MANAGING TECHNOLOGY
DEPENDENT OPERATIONS

Table 1: Evaluating Management Style

Questions of Style Feature Prominence	Low				High
Involved subordinates in decision making	1	2	3	4	5
Shares business information with subordinates	1	2	3	4	5
Strives to develop subordinates	1	2	3	4	5
Encourages subordinates to improve themselves	1	2	3	4	5
Allows subordinate to act independently	1	2	3	4	5
Gives credit to subordinates whenever possible	1	2	3	4	5
Accepts and encourages change	1	2	3	4	5
Is computer literate and encourage everyone to be	1	2	3	4	5
Keeps up with relevant technology	1	2	3	4	5
Constantly evaluates new technology	1	2	3	4	5
Has open communication up, down, and horizontally	1	2	3	4	5
Is more interest in organization than person gain	1	2	3	4	5
Act more as coach then as boss	1	2	3	4	5

These fourteen questions do not provide a definite measure of a manager's style, but rather gives an indication of the managers approach and style. It is suggested that the reader do this evaluation for themselves as well as for their immediate superior. This will provide at least a first order benchmark for the reader within their immediate organization.

It is rare for an organization to be operating throughout and at all levels as an HPO. It is the writer's personal view that it is possible for a unit of an organization to operate in a more HPO fashion than the organization within it is housed. If the reader is more HPO oriented than his or her superior, it may still be possible to operate the organizational unit under their control in an HPO fashion - perhaps not all of the way, but moving in that direction.

There really is no passing or failing score for this profile. Anyone who rates themselves or others as a 5 on all 13 questions is likely a liar. Even the most HPO oriented manager in the world would not, and probably should not score a 5 on every question. As an example, consider the first two questions.

THE MANAGEMENT CHALLENGE
By Keith McKee

If subordinates were involved in every decision the organization would have a "nervous breakdown" and nothing would get done. Actually the good HPO manager might make 80 or 90% of the day-to-day decisions by him or herself because they are simply not important enough to involve the subordinates. At the same time the HPO oriented manager would, whenever possible, involve subordinates in the 10 to 20% of the decision that are critical to the organization and the people in it.

> *Sharing information that is not relevant may cause confusion...*

Sharing business information relative to the organization unit in question is desirable and should be encouraged. If the manager attends a meeting and hears in detail about the "trials and tribulations" of another business unit, sharing with subordinates is likely to have no benefit and could be confusing. Sharing information that is not relevant with the subordinates may cause confusion without benefit. If there is nothing the subordinates can do and if it does not directly impact them, they should not be involved.

Other of the questions really should be answered with a 5 for a manager that is HPO oriented. For example, questions 3, 4, and 6 would seem to fall in this category. The point is that there is no correct answer and no definite score. Hopefully these questions do give an indication of the management style that a person normally uses.

TEAMS

The concept of participation is embedded in all of the organizational forms that we have classified in this chapter as High Performance Organizations. Regardless of the organizational structure, the objective is to give employees at every level greater responsibility and control of their own activities. It should be noted that empowerment does not require being involved in a team. Empowering people can be accomplished in a variety of ways.

MANAGING TECHNOLOGY DEPENDENT OPERATIONS

An individual can be empowered with relatively little formality. For example, a secretary can be asked to handle routine correspondence without involving the boss. In this instance, the secretary prepares, signs, and mails the response without involving her boss. These responses are prepared more rapidly and in less time than if the boss was involved and the boss is free to do other things.

> *An individual can be empowered with relatively little formality.*

A supervisor managing a group of people can operate the group in a participative mode. This means sharing information with the group and soliciting the groups imputes on critical decisions. In a truly participative situation the view of every person in the group will be considered in the decision making process. The decisions made by the supervisor take into consideration the views of the group, but are still made by the supervisor. The group serves in an advisory capacity and hopefully, because they are involved in discussions leading up to a critical decision, will understand and accept the decision even when they are not in total agreement.

The above groups or "teams" are based on the existing organization with the supervisor as team leader and the people within the group as team members. In this instance there is a clear team leader based on the organizational structure. Such teams are long lasting and meet on a regular basis (the staff meeting) or as required for special problems.

Within HPOs, there will also be interdepartmental teams for special problems, for new products, for major system changes, etc. The objective of these teams is to get imput from all of the impacted departments within the organization regarding a common problem. Team members in this situation are representing their home unit and if a specific individual is not able to participate a substitute from the home department can be assigned. Such teams are typically formed by management edict with a team leader assigned. Such teams typically meet as required based on the specific assignment and are disbanded when their mission has been accomplished.

Many organizations are also forming much less formal teams. Under names such as Process Improvement, Quality Improvement, to Worker Participation, etc; teams are organized that have relatively ill-defined goals. Charters for these teams are likely to be broad, such as ideas for improving quality, increasing customer satisfaction, or reducing delivery time. Team members may be assigned or may be volunteers. Coaches or facilitators may be assigned to support these teams, but frequently such teams select their own leaders, recorders, etc.

Operating within broad charters such teams develop a variety of ideas - some small and some major. Management has to evaluate the benefits of these ideas and fund implementation. For other than the smallest projects, the writer would argue that a "person" must be made responsible for implementing the project - a team cannot manage a project.

"Team" has a different meaning depending upon the organization and the context. Three different "team" approaches were discussed above and there are many other variations. Since teams are based on a sports analogy, it is worth noting that "team" has an equally wide range of meaning. The baseball teams organized at the company picnic are "free form" with leadership coming naturally. On the other hand, a basketball team coached by a person who throws chairs and strikes players is also called a team. In the later instance the team is on the court together, but it seems unlikely that

> *...team meaning is to imply that team members are seriously involved in the decision making process*

imput from the individual players has much currency on the strategy used during the game.

For our purposes, team is meant to imply that team members are seriously involved in the decision making process and are treated with respect. A recent book, *Getting Started in Project Management* by Paula Martin and Karen Tate offers a "Sample Team Contract" (5). This "Sample Team Contract" is suggested as a starting point for anyone organizing a team. Clearly this contract has to be modified for each organization. The idea of having a formal document about what is expected of team members and (by inference) the team leaders could be useful for any team. Such a document would include the following:

- **Team Commitments**

- **Team Meeting Rules for**
    - **participation,**
    - **communication,**
    - **problem solving,**
    - **decision making,**
    - **handling conflict.**

- **Meeting Guidelines**

- **Meeting Procedures**

A detail sample Contract is given in the Appendix at the end of the chapter.

TEAMS, PARTICIPATION & PROJECTS

Organizations worldwide are moving toward the High Performance Organization. One of the key drivers for this move is the rapid spread of technology which can only be handled by involving people at all levels throughout the organization. Introduction of new technology requires involvement and decision making at all levels of the organization.

The shift from traditional organizations to HPOs can be complex and challenging. The organization, the management, and the people all come to the process with differing backgrounds and experiences. Earlier we considered organization types, management styles and the changing role of the people. Assuming that most organizations are moving toward HPOs, a very important question is what program management tools are likely to prove most useful. The answer unfortunately is not all that obvious. It depends on where the organization is and where it is going. There is no one model of project management and each organization will pick and choose from available models to select the mixture appropriate for their operations. The following paragraphs comment on the appropriateness of the PM tools currently used in each of the organization types considered earlier.

> *...no one model of project management...*

Project Management Organizations

As was pointed out in considering organizational types, a project can be done without teams and without participative management. The project management tools were developed for and remain most universally used in authoritarian organizations. There is a person with a staff who has both the responsibility and authority to manage the project. The project manager in project based organizations, can and must interact with those people responsible for the various tasks within the project, but typically this is one-on-one interaction. On large projects, these task

> *On large projects ... leaders work for different companies and in different parts of the country*

MANAGING TECHNOLOGY
DEPENDENT OPERATIONS

leaders work for different companies and are located in different parts of the country. Depending on their organization, these lead persons may interact with the people involved on the task as a team or individually. The point here is that concepts like teams and participative management may have little meaning in project based organizations. Force fitting the formal project management tools on projects within HPOs may well be in appropriate. At the same time, there will undoubtedly be situations for which these tools will not be appropriate.

In-House and Contract Service Organizations

The project management tools used for contract and in-house service organizations might be more appropriate for HPOs since they are often used in team type operations. These tools are used in situations where teams of professionals typically work together on smaller projects. The leaders here are technical people who can communicate on a technical level with the team members. The leadership style here is much looser than in project oriented organizations, but typically the team leader is the boss who after the team meeting takes responsibility for the project. The team members participate on the team, but may have little if any decision making power outside of their participation on the team. Project management in these organizational types is based on teams led by technical managers with individual participation beyond the teams limited.

> *...select the "tool" that seems most appropriate...*

The key is to select the PM techniques here would seem most likely to be applicable. Having said that the challenge is that there are no "standard PM tools" for these organizations, various organizations have developed standards for their own systems. Samples of these systems are available and can be considered as models for use in HPOs.

So What Tools Are Appropriate for HPOs?

HPOs are based on broad participation of everyone within the organization with teams organized at many levels to tackle different challenges. Presumable as these teams implement changes there would be "projects." "Projects" would also continue to be initiated by senior management. "Projects" would also continue to be initiated by the service groups within the company. Projects within a HPO will range from simple one-person activities that might be accomplished in hours or days to portfolios of projects that might involve many people over months or years.

There is no single system that will work. HPOs have to pick and chose from project management techniques appropriate to each of their "projects." Management at all levels within the HPO should have an overall knowledge of Managing Technology Dependent Operations so that they can select the "tool" that seems most appropriate from this toolbox.

THE CHALLENGE IS TO SELECT
THE BEST TOOLS
FROM THE TOOLBOX

References

1. International Technology Education Association, *"ITEA/Gallup Poll Reveals What Americans Think About Technology"*, March 2002.
2. John O'Neil, *"On Education and the Economy: a conversation with Marc Tucker", Educational Leadership, March 1992.*
3. Robert Firenze, "Moving Toward a High Performance Organization", The Manufacturing Productivity Center publication, Illinois Institute of Technology, Chicago, 1999.
4. Ekvall, G., Arvonen, J. "The Climate Metaphor in Organizational Theory", M. Bass & P.J.D. Drenth (Eds) Advances in Organizational Psychology (pp 177-190), Beverly Hills, CA, Sage 1987
5. Goran Ekvall of the Swedish Council for Management and Organizational Behavior
6. Paula Martin and Karen Tate ,"Getting Started in Project Management", John Wiley & Sons, 2001

THE MANAGEMENT CHALLANGE
By Keith McKee

APPENDIX

TABLE A-1: SAMPLE TEAM CONTRACT

A. Commitments: As a project team - We will

1. Only agree to do work that we are qualified and capable of doing.
2. Be honest and realistic in planning and reporting project scope, schedule, staffing and cost.
3. Operate in a proactive manner, anticipating potential problems and working to prevent them before they happen.
4. Promptly notify our customer(s) and sponsor of any change that could affect them.
5. Keep other team members informed.
6. Keep proprietary information about our customers in strict confidence.
7. Focus on what is best for the project as a whole.
8. See the project through to completion.

B. Team Meeting Ground Rules: Participation - We will

1. Keep issues that arise in meetings in confidence within the team unless otherwise indicated.
2. Be honest and open during meetings.
3. Encourage a diversity of opinions on all topics.
4. Give everyone the opportunity for equal participation.
5. Be open to new approaches and listen to new ideas.
6. Avoid placing blame when things go wrong. Instead, we will discuss the process and explore how it can be improved.

C. Team Meeting Ground Rules: Communication – We will

1. Seek first to understand, and then to be understood.
2. Be clear and to the point.

MANAGING TECHNOLOGY
DEPENDENT OPERATIONS

      3. Practice active, effective listening skills.
      4. Keep discussions on track.
      5. Use visual means such as drawings, charts and tables to facilitate discussion.

D. Team Meeting Ground Rules: Problem Solving – We will

      1. Encourage everyone to participate.
      2. Encourage all ideas (no criticism) and new concepts from outside our normal perceptions.
      3. Build on each other's ideas.
      4. Use team tools when appropriate to facilitate problem solving.
      5. Whenever possible, use data to assist in problem solving.
      6. Remember that solving problems is a creative process — new ideas and understandings often result.

E. Team Meeting Ground Rules: Decision Making – We will

      1. Make decisions based on data wherever feasible.
      2. Seek to find the needed information or data.
      3. Discuss criteria (cost, time, impact and so on) for making a decision before choosing an option.
      4. Encourage and explore different interpretations of data.
      5. Get input from the entire team before a decision is made.
      6. Discuss concerns with other team members during the team meetings privately rather than with non-team members in inappropriate ways.
      7. Ask all team members if they can support a decision before a decision is made.

F. Team Meeting Ground Rules: Handling Conflict – We will

      1. Regard conflict as normal and as an opportunity for growth.
      2. Seek to understand the interests and desires of each party involved before arriving at answers or solutions.

3. Choose an appropriate time and place to discuss and explore conflict.
4. Listen openly to other points of view.
5. Repeat back to the other person what we understand and ask if it is correct.
6. Acknowledge valid points that the other person has made.
7. State our points of view and our interests in a nonjudgmental and nonattacking manner.
8. Seek to find some common ground for agreement.

G. Meeting Guidelines:

1. Meetings will be held every _days/weeks/months.
2. Meetings will be called by _
3. Agendas will be issued every days/weeks in advance by
4. Meetings will be facilitated by _
5. Evaluations of meetings will be conducted (*after*) every meeting.
6. The scribe will issues minutes within _ days of the meeting.

H. Meeting Procedures:

1. Meetings will begin and end on time.
2. Team members will come to the meetings prepared.
3. Pagers and cell phones will be turned off during meetings.
4. Agenda items for the next meeting will be discussed at the end of each meeting.
5. A Parking Lot will be used to capture off-the-subject ideas and concerns
6. Unresolved issues will be added to the Issues list.
7. If a team member cannot attend a meeting, he or she will send a representative with authority to make decisions.
8. Meeting tasks will be rotated among members.

Chapter 2, Lead-in:

ORGANIZATION AND OPERATIONS - LARGE AND SMALL

This chapter discusses the role of organizations and their operations. It is not a very easy task. Any activity involving more than one person can be considered to be an organization. Size means a lot. For instance, the difference between a multi-divisional, multi-national diversified corporation and an entrepreneurial single-product operation is likely to be in the financial and human resource capability that affects its organizational structure. Structure and management practice will also depend on the purpose and function of the organization.

Operations is a very broadly definable function that is dedicated to getting something done. For a service organization, it may be what goes on in the "back room." In manufacturing, it involves all activities where "there is something that can be dropped at your feet." This "safety shoe" definition of operations is in fact widely accepted and often has a key organizational role at the senior vice president level.

Such "foggy" definitions, while real, are very unsatisfactory. "Operations" in terms of what is expected from its management is presented in the third chapter. There you will find a very concise summary of what traditionally is considered to be the basic components of operations, as can be found in many Operations Management textbooks.

MANAGING TECHNOLOGY
DEPENDENT OPERATIONS

CHAPTER 2

ORGANIZATIONS AND OPERATIONS LARGE AND SMALL

Keith McKee

> *...organization is the "suitable disposition of parts for performing vital functions"*

Organizations and Operations are both commonly used terms that have a wide variety of meanings. The basic definition of organization is the "suitable disposition of parts for performing vital functions." The most common understanding is that an organization is any grouping of people, but an organization doesn't have to have any full time staff. An example of the latter would include social or business clubs – they may be highly structured without any full time staff. For the purposes of this book, it is adequate to consider an organization as a grouping dedicated to conducting operations.

The basic definition of operations is "the act of working or acting to produce the desired result." Operations can refer to the actions taken by surgeons on patients. Operations can also refer to the act of controlling an auto, truck, or other vehicle. Obviously this is not the definition of interest here. For the purposes of this book, operations can simply be considered the function of the organization.

ORGANIZATIONS AND OPERATIONS LARGE AND SMALL
By Keith McKee

> **ORGANIZATIONS ARE GROUPS
> DEDICATED
> TO
> CONDUCTING OPERATIONS
> TO
> ACHIEVE A DESIRED RESULT**

The above definitions are intentionally vague because the contents of this "toolbox," with interpretations, should be applicable to any organization. If there is any single factor that has to be considered in making the interpretations, it is the size of the organization. There are major (and perhaps unavoidable) differences between large and small organizations. A small organization is not simply a small version of a large organization.

A small organization is not simply a small version of a large organization.

This chapter provides further discussions of:

- Organizations
- Operations
- Small versus Large

MANAGING TECHNOLOGY
DEPENDENT OPERATIONS

ORGANIZATIONS

> *...an activity involving more than one person is an organization*

Organizations can serve a wide variety of purposes. Each organization has specific structure and terminology. Every organization provides some sort of product or service to individuals or to other organizations. In a consumer-oriented organization, those receiving the output are referred to as customers. In a law office, they are referred to as clients. In a school, they are referred to as students. In a church they are referred to as parishioners. In a prison, they are referred to as inmates. In a town or city, the are called citizens. The point being made is that every planned activity involving more that one person is an organization. A correlation to that point is that every organization has "customers," whatever they might be called. Every person is involved in a variety of organizations in a variety of roles.

Organizations can be "structured" in a wide variety of ways, but there almost always is some formal structure. Even "pick-up" baseball or basketball games will soon have a person making key decisions. Business organizations, because of the IRS if for no other reason, will always have a formal legal structure. These can vary from Sole Proprietors, to Partnerships, to S Corporations, to C Corporation, to Nonprofit Organization, to Not-for-Profit Organizations, etc. The legal form of the organization is not particularly relevant, but the fact there must be some individual or individuals empowered to act on behalf of the organization should be noted. There is also no relationship between the legal form and the size of the organization. A partnership may be a worldwide organization with tens of thousands of employees while a "C" Corporation can have one employee.

All organizations have structure. The structure organizations can be highly variable. A business with only two partners may operate with both

ORGANIZATIONS AND OPERATIONS LARGE AND SMALL
By Keith McKee

partners having equal power and all decision being made jointly. In larger organizations, including partnerships, there will always be a formally established management structure – in the case of a partnership a managing partner with other partners in charges of offices, practices, etc. When a Sole Proprietorship is small the top manager may simply be the owner, but as the organizations grows there will typically be all of the formal organizational structure starting with CEO, CFO, CTO, President, etc. etc.

> *All organizations have structure.*

Organizational structure also depends on the ownership and purpose of the organization. A privately held for-profit business will be different than a publicly held for-profit business. For a privately held enterprise, the primary objective of the owners may be to increase the value of the enterprise by maximizing growth and minimizing profits, which are taxed. A publicly held for-profit business operates to maximize the growth of stock price. Not-for and non-profit organizations, as well as government agencies at all levels, can have a wide variety of objectives where typically growth and/or profits are not involved. The author would argue that an organization exists to serve at least four constituencies:

1. The stakeholders who own or in some way control the enterprise
2. The customers or clients of the enterprise
3. The employees and/or members of the enterprise
4. The general public

Some people would argue that all organizations also exist to serve the government – certainly all firms have to be legally acknowledged by the government in order to exist. Almost all organizations also have to submit some sort of reports to one or more governmental bodies.

Organizational structures can be based on a variety of models. In a subsequent chapter, teams are considered as one tool for use within an organization. The organization of a 'high-tech start-up" company is typically

MANAGING TECHNOLOGY
DEPENDENT OPERATIONS

Organizational structures can be based on a variety of models little more than a company-wide team. Each person on the team is fully integrated into the organization with the individuals working together for the benefit of the overall organization. In this instance the team leader may be the person starting the company or their representative, but all of the other workers are team members contributing as individual to maximize the overall results. The author would argue that for a team to produce any significant results there must be a leader. In some circumstances the team may select a leader, but normally that leader comes with some additional powers based on his/her position within the organization.

On the other extreme, the organizational structure that most often comes to mind is the hierarchical structure based on a military model. A sample of such an organization chart was included as Figure 2 in Chapter 1. In this type of organization there is a clearly defined reporting structure with everyone knowing exactly who reports to whom. Relatively small groups of individuals (typically 5 to 10) within the organization report to a single person at the lower management level. A relatively small group of those managers (again typically 5 to 10) report to the next higher level of management. In such a hierarchical organization, the lines for reporting are completely defined and there is a single person on top of the pyramid directing the entire organization.

The hierarchical structure is "neat and clean" with the lines of authority always perfectly clear. Within a given unit, the manager is "the boss" and the direct reports work for him or her and communicate up only through "the boss." If the workers within the group need to communicate with a parallel unit, that communication, at least in theory, goes through "the boss" to "the boss" of a parallel organizational unit. In this model, workers within a unit have no need to know about or be concerned with what parallel units or higher units are doing. In theory any communication required between units goes through the boss who communicates with the boss of the second unit.

ORGANIZATIONS AND OPERATIONS LARGE AND SMALL
By Keith McKee

Many of the modern management techniques are attempting to open communications between units in hierarchical organizations, but with very few exceptions the military model is still the norm.

> *...the attempt is to improve, not replace, the traditional organization.*

High Performance Organizations, World Class Organizations, TQM, Six-Sigma, and the like basically represent fixes laid upon the traditional hierarchical organization – not an alternate structure. If such an alternate structure exists, it is not obvious to this writer. This is an important point – the attempt is to improve, not replace, the traditional organization. Within the hierarchical organization, the people in "top management" may work as a "management team", i.e. the key units managers of the business work together as a "team" to arrive at overall decisions. For example; the President, VP of Marketing/Sales, the VP of Operations, and the VP of Finance/Business will jointly work on corporate-wide problems while at the same time managing their individual corporate units. On such a management team, it is clear that the team leader, i.e. the president, has the final decision making power.

As mentioned in Chapter 1, some organizations are project based. Such organizations are hierarchical with a president and appropriate VPs. The difference here is that the operating portion of the business is based on projects. The primary assignment of employees within such an organization is to a project, which may last for years. When that ends, he or she is reassigned to a different project. Clearly the advantage is that the reporting lines are clear since the employee is working on a given project until he or she is reassigned to a different project. The disadvantage is that individuals do not necessarily work with technical peers; they work full time with others assigned to the project. One negative results of this organizational structure is that individual specialists may have difficulty keeping current in their technical areas.

MANAGING TECHNOLOGY DEPENDENT OPERATIONS

> *...typically employees are assigned to business units based on their professional expertise*

Organizations of all types may make a specific large, long-term project the full-time assignment for one or several employees. This, however, is the exception. When projects are relatively small or of relatively short duration, "disorganization" results from assigning people full time to specific projects for short periods. More typically, employees are permanently assigned to groups based on their professional expertise, e.g. marketing, engineering, or manufacturing. Employees are then assigned to teams based on availability as well as the need of the individual teams. Under such circumstances, the employee normally works in a group of peers within the "home" group and with other professionals on the project. The team assignments are temporary, part-time activities that are normally external to the "home" group. This matrix organization is illustrated in Figure 1.

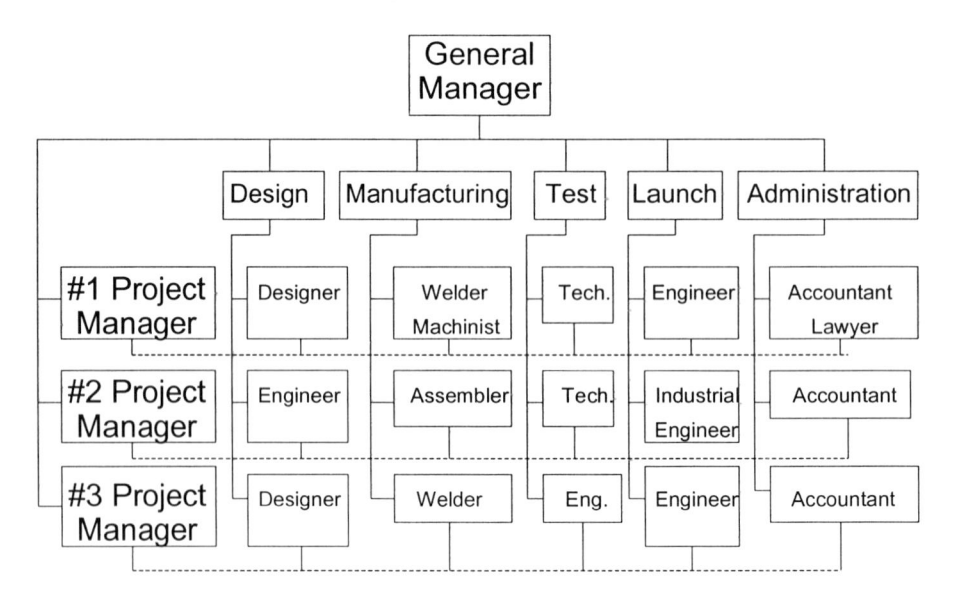

FIGURE 1: MATRIX ORGANIZATION

	Design	Manufacturing	Test	Launch	Administration
#1 Project Manager	Designer	Welder Machinist	Tech.	Engineer	Accountant Lawyer
#2 Project Manager	Engineer	Assembler	Tech.	Industrial Engineer	Accountant
#3 Project Manager	Designer	Welder	Eng.	Engineer	Accountant

All reporting to General Manager.

ORGANIZATIONS AND OPERATIONS LARGE AND SMALL
By Keith McKee

The challenge of the matrix organization can be significant. Since raises and promotions come from the manager of the "home" group, the individual worker assigned to an external team has divided loyalties and often gets mixed signals.

Teams clearly can operate internally to a "home" group, which is relatively simple since the person responsible for the team is also the manger of the "home" group.

> *the most valuable teams are cross-discipline teams*

As is discussed in other chapters, the most valuable teams are cross-discipline teams bringing together people representing different backgrounds and experiences. When a staff member is assigned to a team outside of the "home group" there are a variety of potential conflicts. The "home" group manager and the employee assigned to the external teams are both challenged. The manager has lost what may well be one of the most valuable people for some period of time or at least for part of their time. Unless this is taken into account when assigning work to the "home" group, the result is likely to be conflict – open or even worse hidden. The employee assigned to a cross functional team, may find him or her self badly conflicted – required to work for two masters who both want the same time and dedication.

Teams organized within a "home" group clearly have the least complexity because the permanent assignment and the team activities are being done under the same manager. Projects organized within broader functions, e.g. engineering or manufacturing, have some conflict built-in, but since the common management is normally within sight this is not usually a major issues (or at least management can clearly see and resolve problems that arise.) Truly cross-functional teams that involve people from different business units, different locations, and difference functional areas represent the major challenge. Such "matrix" assignment of people is mandatory to setup effective project teams for the most serious problems facing the organization. At the same time, this is the organizational structure that requires the most care and attention. For matrix management to work, it

MANAGING TECHNOLOGY
DEPENDENT OPERATIONS

must be strongly supported by "top management" and the willing cooperation of individuals and managers throughout the organization.

> **EVERY ORGANIZATION HAS A STRUCTURE**
> **THERE HAS TO BE SOMEONE IN CHARGE**
> **OR NOTHING WILL GET DONE**

OPERATIONS

The title of this book contains the word "operations" used in a very generic fashion – operations could be the function of any organization or organizational subset that is dedicated to getting something done. Perhaps the operative words in the book title are "technology dependent." Based on that interpretation, the definition of operations is totally open and can be considered as:

> **OPERATIONS ARE THE PURPOSE**
> **FOR WHICH THE ORGANIZATION EXISTS**

This very broad definition would mean that the operation of the Accounting Department would be to do the accounting. The operation of the Manufacturing Department would be production of the product. The operation of the Marketing and Sales Department would be obtaining and retaining customers. Such examples might either appear to be overly simplistic or evasive. Since such "foggy" definitions are very unsatisfactory, "operations" is considered here in greater details.

Perhaps the most common use of the word "operations" is that portion of an organization that does the actual work. In the case of a manufacturing

ORGANIZATIONS AND OPERATIONS LARGE AND SMALL
By Keith McKee

enterprise, operations will always include the actual manufacturing. In a distribution/warehouse enterprise, operations would include all of the handling and storing of the materials. One definition of these

> ..."operations" is that portion of an organization that does the actual work

operations are that they involve activities where "there is something involved than can be dropped on your feet." This "safety shoe" definition of operations is in fact widely accepted and many companies have a VP of operations responsible for this type of activity and other VPs responsible for what is done in the balance of the company.

Within service enterprises, operations refers to the "backroom activities", e.g. for a bank it would include all of those activities normally done out-of-sight of the customers. Within a software company, it would be those people actually writing the code that becomes the software. In a RD&D operation it would be the engineers and scientist doing their jobs. There may be different titles, but again there frequently is a person make responsible for something called operations.

Operations as defined in most textbooks would typically include operations as discussed in the previous two paragraphs. The two key portions of an enterprise that are not normally

> ...actual reporting structure may trace it roots to history, personality, or chance.

included as part of operations are Marketing/Sales and Finance/Business. In even the most elementary enterprises, there will normally be at least these three major activities: Operations, Marketing/Sales, and Finance/Business. Even small enterprises where one person may do more than one of these functions, there will still be a person with clear responsibility for each of these activities.

MANAGING TECHNOLOGY
DEPENDENT OPERATIONS

> *...actual reporting structure may trace its roots to history, personality, or chance.*

Depending upon the organization, there are a number of activities that may operate as stand alone or may report on any of the three operations listed above. Examples here could include IT, Quality, Purchasing, Distribution, Engineering, R,D,&E, etc. The actual reporting structure may trace it roots to history, personality, or chance. The organization charts are clearly critical to the people involved, but to a significant extent they confuse the issue. Each of the organizations within a company performs some service for the operations. For the purposes of this book any organization or sub-organization that is "technical dependent" is included. Based on this we would argue that since there is no portion of a business enterprise that is not being impacted by change, every organization has to learn how to deal with new technology.

Based on this logic, the authors believe that the contents of this toolbox are useable by anyone responsible for supervision or management of any operation. In Chapter 3 the traditional topics of operations management are discussed in greater detail.

LARGE VERSUS SMALL

The technical challenges faced by large or small organizations are very similar, but the resources available to handle them may be totally different. A large organization may have a Chief Technical Officer (CTO) with appropriate education and staff to understand and evaluate even the most complex technologies. Within a small organization, the same function is likely to be filled by the person in charge of the organization or the person in charge of operations. In either instance that person will likely have very limited technical background and no technical support staff. It is very important to keep these differences in mind:

ORGANIZATIONS AND OPERATIONS LARGE AND SMALL
By Keith McKee

**A SMALL ORGANIZATION
IS NOT
SIMPLY A SMALL VERSION OF A LARGE ORGANIZATION**

This sounds logical and the reader is likely to be saying that everyone knows that. Unfortunately although that observation may seem obvious, over the years, the author has noted that most business textbooks, most business case studies, and almost all articles about business talk about the large companies. Most business textbooks would make it seem that a small company is one with only a few hundred workers. Rarely, if ever, are there examples or case studies that seriously consider smaller businesses.

Part of the problem may be that the Federal Government defined a small business as one with less that 500 employees. A 500 person company is small compared to a company with many thousands of employees but looks gigantic to someone involved with a company that has 10, 20, or 50 employees. A 500-person company is large enough to have all of the structure that we associate with large companies. Such organizations may well be simply smaller versions of large companies. For the purposes of this book, it is suggested that a small company as one that is too small to have the organization and infrastructure of a large company.

A small organization is normally directed and likely owned by the person in the corner office. The owner and perhaps 2 or 3 trusted lieutenants handle all of the management functions. Each person may "wear several hats," and often these key people handle multiple functions. For example, the owner may

Approaches that may be logical and rational for a large organization may be impossible for a small organization

MANAGING TECHNOLOGY
DEPENDENT OPERATIONS

be the salesman, the chief financial officer, and the engineering manager in addition to being the president and owner. There is an organization, but it looks nothing like the formal assignment of duties that one expects in a larger organization.

Approaches that are logical and rational for large organizations may be impossible for small organizations. For example, large organizations may have project management specialist who have the latest computerized tools for project scheduling and management. The small organization may have to depend upon the project managers operating with limited support where a manually prepared Gantt chart is the level of technology available. Similarly, the large organization may have a group of scientist and engineers who are available when technical issues are encountered. Within the small organization, the same person who is drawing the Gantt chart may also be the only technical expertise.

> **RESOURCES WITHIN A SMALL COMPANY MAY BE TOTALLY DIFFERENT FROM THOSE WIHIN A LARGE COMPANY**

The problems are similar, but the available staff and tools are totally different.

...operations are or are becoming technology dependent........

The difference between large and small organizations becomes less significant as one gets closer to front line employees. The challenge of involving people within in a department of a large organization is the same as within a small organization. Basically people are people and the challenge of getting them involved in evaluating and implementing new technologies is not significantly different.

ORGANIZATIONS AND OPERATIONS LARGE AND SMALL
By Keith McKee

This toolbox has been prepared so that it should have value to managers within both large and small organizations who are involved in management of any type of operation. As observed earlier, it is the author's view that all operations are, or are becoming, technology dependent. Because of the potential difference between large and small organizations, the editors have added occasional notes to provide "translations" between these two situations.

Chapter 3, Lead-in:

WHAT CAN BE EXPECTED OF OPERATIONS MANAGEMENT?

As already suggested, the definition of operations can be taken broadly. Even when defined by a formal organizational structure, such as Operations or Production & Operations that is defined more narrowly, (i.e. distinct from Sales, HR, or the Legal Department) its interaction with these functions make its management a very far-reaching endeavor. It is the core of the value added by any enterprise. It integrates technology applications into the creation of the products that bring in the revenue. It is the heart and life-blood of a business. It is what makes the thing run.

So, what are the basic elements of operations management that a non-technical manager, given the responsibility, may have to deal with?

The essence of this chapter is, therefore, to briefly introduce the topics an operations management textbook would teach. It is a capsule abstract of what a course in operations management would cover in an MBA program at a management school. To a manager of technology dependent operations it should point out what under his or her circumstances needs to be pursued in greater depth. References are provided for this purpose. In addition, Chapter 12 brings additional insights of the most recent and technologically specific operations issues from the practitioner's point of view.

MANAGING TECHNOLOGY
DEPENDENT OPERATIONS

Chapter 3

WHAT CAN BE EXPECTED OF OPERATIONS MANAGEMENT?

D. Tijunelis

To appreciate the challenges and find tools to manage technology dependent operations it is necessary to have an understanding of what constitutes operations management. Nevertheless, there are situations where one gets assigned to manage operations without any formal operations management training. This chapter is an introduction to a broad overview. Follow-up is left to the reader.

Operations Management (OM) is a core subject in all management programs. OM covers both quantitative and qualitative methods. Generally it is one of the more difficult courses because it covers such a wide range of topics of business administration or MBA programs. The following topics are typically included under Operations Management:

WHAT CAN BE EXPECTED OF OPERATIONS MANAGEMENT?
By D. Tijunelis

- Operations strategy
- Project management
- Process analysis and job design
- Product design and process selection
- Quality management
- Supply chain strategy
- Forecasting
- Resource planning and control
- Inventory management
- Operations scheduling
- Regulations: Safety & environment protection

Operations management provides a systematic means of observing organizational processes. It has elements of organizational behavior as well as hard science and technology. It applies to manufacturing as well as service enterprises. It provides managers with tools that can be applied in a variety of jobs and industries. Operations management focuses on the core conversion processes of the firm, be it a manufacturing or service business. This is where value is added for the customer. It is therefore the key function of any enterprise. The following introduces the key topics in brief.

> *This is where value is added for the customer.*

• Operations Strategy

Operations are those activties where value is added for the customer. Strategic considerations in operations management include capacity planning, facility planning, technology planning, workforce development, quality planning, production planning, and workplace organization. In *Manufacturing Strategy* by Terry Hill (1) there are suggestions for an approach by which companies can identify order-winning criteria in an effort to obtain effective operations strategies. The Malcolm Baldrige National Quality Award program facilitates comprehensive benchmarking of the effect of operations strategy.

MANAGING TECHNOLOGY
DEPENDENT OPERATIONS

It requires the operations strategy to fit the corporate strategy. The operations manager has to address the core competitive priority issues of

- cost,

- quality,

- timing,

- flexibility.

> *Core competency can result in core rigidity - ...*

Distinctive competency assessment is a key part of operations strategy development. This is not easy because there is a lot of bias. Traditionally, operations management is the most stable part of an enterprise. After all, operations are expected to provide reproducibility and low cost. The learning curve taught in schools suggests that for each doubling of cumulative volume of production of the same product, its unit cost should go down by 10-20%. However, the learning curve approach encourages repetition and, as a result, encourages resistance to change rather than flexibility. Core competency can result in core rigidity – shortsighted unwillingness to change and the loss of strategic flexibility. A classical example was Ford's Model T. For a while it was a very successful application of the learning curve. However, Ford did not recognize its limits in time. It resulted in the costly, almost catastrophic, problem of converting operations into cost effective production of the Model A and subsequent models until the innovations that resulted in the Mustang in the 1960's.

One of the basic elements of operational strategy is a continuum of positioning with respect to the product or process life cycle - the length of time a product is in demand or the time that a process is superior to alternative means of production. At what point in time are operations expected to shift from custom products to highly standardized products? This will involve major plant layout considerations. Optimizing these can get complex and computer simulation may be needed. Capacity planning, break-even analysis and decision-trees are common tools. The positioning strategy between low volume and high volume production is obviously dependent on a high degree of coordination between marketing and production/operations staff – a leadership issue. Consider the audit questions of Table 1.

Table 1: Operations Strategy Audit

Audit Questions:		Core Competencies			
		Efficiency	Quality	Customer Response	Flexibility
Are we	good				
	average				
	poor				
Are we better than competition?					
Where can we be the first to improve?					
How?					
Who controls?					

Operations strategy is an essential component of almost every article, book, and textbook dealing with strategic management. Beyond the reference mentioned above, *Strategic Management: An Integrated Approach* by Hill and Jones (2) is a classic that puts operations strategy issues in the broader context of corporate and business strategies.

MANAGING TECHNOLOGY
DEPENDENT OPERATIONS

Project Management

> ...*putting the maximum of their resources on targeted goals*

Many businesses are trying to be as flexible as possible, putting the maximum of their resources on targeted goals. They are doing it with ad hoc organizations. Pooling the resources by project is one way of accomplishing this. Many efforts are undertaken by project, or a group of projects, as a focused program to change capacity, improve quality, employ new technology and so on. Project management differs from traditional management, in purpose, structure, and operation. It deals with work breakdown structure and critical path methods (CPM), time/cost models, and limited resource models. Project managers discuss time/cost trade-off using networks or project task sequencing like PERT/cost monitoring techniques. For limited resources, they consider problems of workforce leveling and resource allocation.

Under the project management heading, besides specific task scheduling and control programs, operations textbooks teach the use of project logic. Are your operation's project's objectives SMART? (3) (See Table 2) A specific discussion of tools for project management is available in Chapter 11.

Table 2: Smart Operations Project Audit

SMART	Objective	yes	no	?
Strategic	to serve business goals			
Measurable	to have clear deliverables			
Achievable	to be technically doable			
Realistic	to be practically doable			
Timed	to be time limited			

WHAT CAN BE EXPECTED OF OPERATIONS MANAGEMENT?
By D. Tijunelis

Process Analysis and Job Design

This topic covers the basic process flowcharting, process analysis, and the application of current trends in job design. Some of these trends include implementing quality control in jobs, cross-training workers, employee involvement, team involvement, information technology's impact on work, use of temporary workers, automation of heavy manual work, and increased emphasis on job satisfaction. These trends place pressure on management to pay close attention to job enrichment, worker interaction with machinery, and interaction with coworkers.

> ...*pressure on management to pay close attention to job enrichment...*

Work flowcharts, using arrows for the sequence of movements between rectangles for transformation tasks, triangles for storage, and diamonds for decision points, describe any process. This simple approach helps analyze an overall process for possible bottlenecks and improvement potential. Obviously, such flowcharts can be very complicated for multi-part or multi-ingredient process with a sequence of many physical transformations and processing options. Figure 1 is an example of the concept.

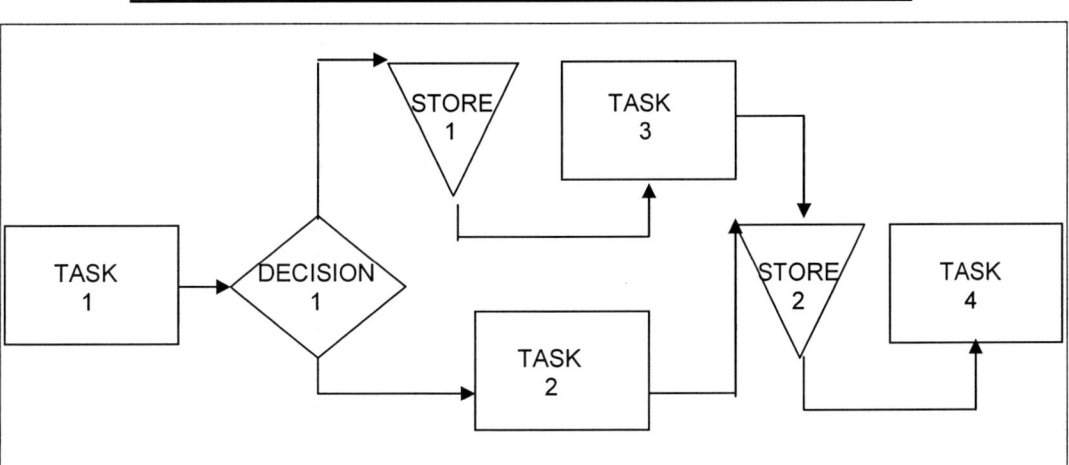

Figure 1: Components to Build a Process Flowchart

MANAGING TECHNOLOGY
DEPENDENT OPERATIONS

The most complicated assembly or chemical processes to be broken down into these simple flowchart stages. It can help you appreciate the work in progress (WIP) and finished goods management issues. (See Table 3) Workflow design is interdependent with the specifics of job design. Job design includes work sampling, work measurement, and time studies. Job enrichment opportunities are identified. Employee health and safety issues are addressed here or are given additional attention as a separate topic.

Table 3: Process & Job Design Audit

Analysis Issues	yes	no	?	Action Needed
Do you understand the sequence and function of the process stages?				
Have bottlenecks been identified?				
Have key jobs been designed with the workflow in mind?				
Have improvement opportunities been identified?				

As a follow-up of techniques for process analysis, refer to traditional operations management textbooks such as those by Gaither (4); Heizer and Render (5); and Chase, Aqilano, and Jacobs Schroeder (6). Job design related discussion can be found in several of the following chapters of this book.

Product Design and Process Selection

...Another important concept has been the adoption of "design teams" and "concurrent engineering."

In recent years, product development has changed in important ways as the customer has been made part of the design process. End products are being designed with manufacturability, maintainability, realiability, and recyclying in mind. Another important concept has been the adoption of "design teams" and "concurrent engineering." Important to management are also the quality of management issues dealing with the team approach, process engineers, product design engineers, marketers, customers, suppliers, and other stakeholders who are simultaneously involved in designing products to be effectively delivered as products or services. At least in part this task falls in the purview of value engineering (VE). VE is a discipline of product optimization to perform all of its needed functions with the minimum of cost and parts for ease of manufacturing.

A problem faced by every company is designing an effective process layout. As business requirements change, layouts are subsequently changed. This applies to layouts for new facilities and existing facilities for both service and manufacturing companies. Process flow designs use group technology (GT) as common manufacturing layout options. In the GT approach, parts are identified (coded) and process equipment arranged for optimized routing to minimize set up and material handling time. No particular type of layout is inherently good or bad, and layouts are often reflective of the organizational makeup of individual firms. There are a variety of simulation tools available to help optimize plant layout.

The contemporary view of services is that the customer is the focal point of all actions in a service organization. This is true for manufacturing as well, only with a less intimate and immediate link. For some sevices, such as in the operation room of a hospital, the actions taken by the doctor have an immediate effect on the customer.

> *Service strategy begins by selecting an operations focus*

There is no distributor, and no lead time. The product of the process may have to be customized on the spot. Accordingly, process design must have built-in flexibility. Service strategy begins by selecting an operations focus, such as treatment of the customer, speed of delivery, price, variety, quality or unique skills that constitute a services offering.

MANAGING TECHNOLOGY
DEPENDENT OPERATIONS

For the case of service business, the objectives for technical operations staff are to model and optimize waiting-line situations. For simple waiting line situations there are standard formulas; for complex waiting situations there is sophisticated software. These take into account arrival time, distribution of arrivals, arrival patterns, time required per arriving unit, and degree of tolerance for the waiting time as well as several other specific factors. It is a classical subject in statistics, probability theory, industrial engineering, and oprations research. Consider asking questions from Table 4.

Table 4: Optimizing Process & Product (Service)

Qestions to Consider	yes	no	?
Is product (service) and process design taking place concurrently?			
Is product optimized through value engineering?			
Is production facility layout being optimized through simulation?			

Quality Management

...product and service quality have become the basis for competition.

A basic understanding of the principals of quality management are essential for any business as it follows a continous improvement and total quality management (TQM) model. This includes the lessons from quality management experts like Edward Deming. He was the quality mangement leader whose advice put the Japanese industry on the path of quality and productivity. His continuous improvement model applications have been generally acceped throughout the industry. Attention to productivity improvement and TQM were promoted by US Secretary of Commerce Malcolm Baldrige. The National Awards system, which provides benchmarking opportunity, was named in his honor. Furthermore, there is also the ISO 9000 system of standards agreed upon by International Organization for Standerdization for quality standards and certification for international trade.

In a competitive environment, quality advantage is very important. In mature or declining industry, it may be the only advantage. Your technical staff will deal with standard quality control (QC) concepts of cost of quality and statistical process control (SPC). This would include Taguchi methodology, as well as broad issues of process design and service quality measurements. Modern treatment of quality considers three levels of industrial quality:

1. The quality of organization

2. The quality of the processes used

3. The quality of the actual performance of the tasks

Between technical operations and other functions, the definition of quality may vary. Manufacturing considers meeting written specifications as the principal criteria of quality. On the other hand, Sales would look at it on a value-based or use-based criteria - that is, fittnes for use or the use potential per unit price. Marketing and R&D may consider quality in judgmental or product definition terms - that is, the one will rely on survey opinions while the other on their own definition of product quality purely based on the exellence of the technical features. (See Table 5)

Table 5: Review of Total Quality Management

Quality Management Questions	yes	no	?
Is quality perception appropriate in each of your organizational functions?			
Is there a coordinated effort toward the ultimate goal of customer satisfaction?			

MANAGING TECHNOLOGY DEPENDENT OPERATIONS

Note, Quality Control and Manufacturing staff will focus on the "manufacturing definition of quality," i.e. meeting the specifications. In doing so, they will need tools such as statistical process control (SPC) charts, cause-and-effect diagrams. There are a wide variety of tools and techniques that have been developed and used for Quality Control. Some of these are Fishbone diagrams, Pareto graphs, Taguchi acceptance criteria, six-sigma approach, inspection frequency charts and statistically designed analysis of variance (ANOVA) experiments. For non-technical management this may sound challenging, but the basic concepts are simple. The book *Management and Control of Quality* by James Evans and William Lindsay does a good job making it easy to understand, while being comprehensive (7).

• Supply Chain Strategy

Supply-chain management traditionally has had the role of the purchasing and shipping departments of a firm. A more current broad definition of supply-chain is the linking of functions that source raw materials, integrates their transformation into products, and distributes these to the customers. This can involve transportation, warehousing, order fulfillment, forecasting, forms of accounting, credit transfers, etc.

> *... supply-chain strategy is critical to operations management.*

In the highly competitive globalizing economy, a supply-chain strategy is critical to operations management. Management must understand the supply-chain dynamics that are inherent in its operations. Here the measure of inventory turns and days-of-supply are important. Topics such as the "bullwhip" or the "wave-swell" effects magnify customer demand back down the chain of supply. It is like the swelling of waves driving into the shore. The order spikes at the consumer level progressively exaggerate, or at least distort, the requests for supply. It is the result of time delay in data gathering, poor forecasting and risk aversion to either build or deplete inventory safety-stock. These affect raw material, outsourcing options and the elements of conventional materials management system, purchasing, intra-plant logistics, and finished goods distribution. Here operations management is impacted by the use of the Internet and associated information technologies. An integrated approach to supply-chain management is important to link together traditional organization functions. Internet infrastructure and business-to-business (B2B) applications should evolve into a new paradigm for operations management. This must include global sourcing, which unfortunately gets complicated by geopolitical issues. Political factors are insufficiently addressed in many operations training programs and must be supplemented by individual awareness of current events through popular media.

Beyond the general rules given in operations management textbooks, there are very few references on the subject of Internet based B2B supply chain management that are timely. This is an evolving function of operations. It has global implications that involve not only political-economy concerns, but also exchange rate factors that are subject to change without much notice. Following current publications, such as Business Week, Industry Week and Purchasing Manager's Index available through National Association of Purchasing Management (http://www.napm.org), is good advice. (See Table 6).

Table 6: Supply Chain Audit

Management Questions	yes	no	?
Is there a focus on supply chain effect?			
Is supplier-link and data timely / accurate?			
Is the risk aversion appropriate?			
Is your global-reach well informed?			

MANAGING TECHNOLOGY DEPENDENT OPERATIONS

· Forecasting

> *Forecasting is a prerequisite to any type of planning*

Forecasting is a prerequisite to any type of planning and is an integral part of production planning and inventory control. For operations management there are an array of techniques available. They are qualitative and quantitative such as time series analysis, causal relationships, and simulation models. Many of the qualitative techniques are employed in market research while the more quantitative are common in forecasting operations demand and resource availability. Such tools are an integral part of the discipline of operatrions research. These days, it is of relatively little value for management to learn the mathematical roots of the forecasts. Since developments are expanding the capabilities of software for forecasting as a competitive tool, it is imparative for operations management to have an appreciation of the application potential and to communicate with the computer competent specialists with respect to practical implementation.

The following Table 7 gives the reader an indication of the forecasting tools and techniques that are available and in use.

Table 7: Forecasting Tools

TECHNIQUES	FEATURE	CONCERN
Qualitative:		
Brainstorming	Uninhibited group input of ideas	Results dependent on facilitator's leadership skill
Nominal Group Technique	Group process with forced secret voting	Difficult to prioritize the independent input
Surveys	Select group formal & structured inputs	Costly and depends on accuracy of questions
Historical data (life-cycle)	Analogies to past events	High relevance and current event dependent
Delphi Method	Process of formal query of experts	Costly, time consuming, and facilitator skill dependent
Quantitative*:		
Time Series	Extend average of data vs. time	Totally dependent on past history
Moving Average	Extend average of more recent data	Allows some judgment of the relevance of current events
Exponential Smoothing	Eliminates extremes of moving average	Depends on subjective judgment of typical behavior
Causal Correlation	Relies on cause and effect relationships	Illusion of precision can obscure false relationship.

*All are available on operations related computer software. *Operations Research and Management Science* (ORMS) publishes an extensive list of software (8).

In operations the mechanics of forecasting is mostly left to computer software. It is, however, important to appreciate the distinction between precision of the forecast and the relevance and reliability of data on which it is based. For longer-range forecasting, a combination of future scenarios based on qualitative forecasts and quantitative extrapolation is recommended.

Resource Planning and Controlling

Capacity-related decisions could either enhance or worsen competitiveness over the long term. Decisions for the best operating level

MANAGING TECHNOLOGY DEPENDENT OPERATIONS

and economies of scale use the techniques of capacity planning process and decison trees. Capacity planning is based on break-even analysis, while decision-trees are a form of logic supported by economic value analysis of options and their probabilities. As operations globalize, location decisions become more complex, and work-force diversity and culture issues rise that are unfamiliar to the traditional technitian.

Material Requirements Planning (MRP) or Enterprise Requirement Planning (ERP) are key to the practice of inventory planning and control for product companies.

Enterprise resource planning (ERP) incorporates the control elements in a broader context for operations manager

Another important practice is just-in-time (JIT) which strives to have parts arrive just in time for the next operation. JIT is centered on design elements such as focused factory networks, group technology and quality at the source, uniform plant loading, and setup-time reduction. JIT concepts are also applicable to services where the waiting-line factors are considered in process design. It is to minimize upstream loading, i.e. like calling to make reservations, or the "who is next" call at the doctor's office.

MRP/ERP are computer-based control systems that monitor activities within the operations in a broad context of the operating systems integration. It depends on computer applications, information systems management and the discipline of information technology (IT). The subject is complex in the state of continual improvement and requires specialists for effective implementation. As an example, the SAP AG corporation has set new standards in information technology with a comprehensive set of applications software. Its significant feature is its use of a data warehouse. Such software packages allow data to be easily aggregated and disaggregated by the user into major components such as financial accounting, human resources, manufacturing and logistics, and sales and distribution.

Another part of the operations planning process is aggregate planning. It is broad and takes in long range capacity decisions and involves personnel issues such as staffing levels. Experience, age, education, and cultural diversiy can become an issue in the communications between the general management, operations management, and the technical specialists.

While the subject of resource planning and controlling is very important to a business operation, it is in a constant state of innovation. Since it is information technology dependent and is continuously being upgraded, up-to-date information is likely to come not from textbook, but rather from industry publications, software developers or independent consultants. (See Table 8).

Table 8: Resource Planning Review

Basic Questions	yes	no	?
Is resource planning intergrated into the strategic planning process effectively?			
Are resource planning tools being deployed efficiently?			
Are tools subject to planned continuous improvement ?			

MANAGING TECHNOLOGY DEPENDENT OPERATIONS

Inventory Management

> *Operations management is responsible for inventory...*

Very often inventory management gets special attention. It is not unusual that more than a third of the investments of a manufacturing firm is tied up in inventory. Operations management is responsible for inventory. Different inventory models are used - those that are based on ordering at set time intervals and those that are based on ordering set amounts as the need arises. Relevant inventory costs consist of holding, setup, ordering, and shortage costs. In today's outsourcing environment, inventory analysis must incorporate supply chain issues in addressing various service levels – the order quantities and reorder points needed to meet demand directly from stock on hand. However, good inventory management should be well interconnected with what the customer wants on one end and with what suppliers are ready to deliver on the other. Ideally the supplier is preparing to bring raw materials just in time to produce what it takes to meet customer wants. Organizationally, the strategic aspects of the inventory – supply chain interrelationship - may be considered as logistics. (See Table 9)

Table 9: Inventory Issues

Questions to consider:	yes	no	?
Is your inventory costly?			
Is your inventory perishable?			
Is your inventory interconnected to the customers?			
Is your inventory interconnected to suppliers?			
Are you connecting your suppliers to your customers?			

WHAT CAN BE EXPECTED OF OPERATIONS MANAGEMENT?
By D. Tijunelis

A number of the above questions are technology dependent. Some simply depend on good management information system (MIS) while others are hardware related, such as the use of bar codes, electronic transponders to automatically monitor shipment movements on the loading dock, etc.

Operations Scheduling

Scheduling drives the operations management engine. In small organizations, it is handled by practical logic and habit. In large companies, it is an integral part of systems integration software. Operations management in manufacturing and service industries often give major emphasis to short-term scheduling. Their methods contain basic rules for shop floor control, input-output control, and capacity planning for manufacturing. In the service area, personnel scheduling is the prime subject. Job-shop scheduling needs priority rules to decide which job to do next based on identified bottlenecks. A bottleneck can be viewed as a resource whose capacity is less than the demand, while all else has excess capacity that can be put to use if the bottleneck is corrected. This leads into synchronous manufacturing and the theory of constraints advanced by Eli Goldratt. To get started on a road to improvement, consider reading or at least skimming E. Goldratt and Jay Cox's book *The Goal* (9). In a general way, examine an overall process workflow chart, separate tasks, decision points, movement options, and work in process as well as final goods storage. Look for possible bottlenecks and direct scheduling to focus on these.

> *Computer simulation technology is used to evaluate priority rules*

One of the most meaningful tools of operations management is benchmarking the operation against an example of excellence. It does not have to be a direct competitor or even in the same industry. It simply has to be relevant as a comparison of a generic function that apparently can be managed better. How to get the comparison is up to you.

MANAGING TECHNOLOGY
DEPENDENT OPERATIONS

- ## Regulations: Safety & environment protection

Public safety is a growing concern. Some books introduce the subject and its significance with regard to the Occupational Safety and Health Administration Act (OSHA) regulations when discussing job and facilities design. Others skim through it elsewhere as it relates to strategy, and quality of consulting practices. It is a subject of considerable technical depth. *Recognition of Technical Hazards in Industry* by Burgess (10) is a good starting reference. Of course, there are government agencies that will provide specific information: Environmental Protection Agency (EPA), 401 M Street, SW, Washington, DC 20460; Air and Water Programs Office; National Institute of Occupational Safety and Health (NIOSH); Publications and Disseminations, DSDTT, 4676 Columbus Parkway, Cincinnati, OH 45226; and The Department of Labor, Occupational Safety and Health Administration, 2000 Constitution Ave., Washington, DC, 20210.

- ## Summing up:

In summing up an overview of Operations Management, consider the questions raised in Tables 10 and 11. They relate to what typically can be expected from an in operations management training program. Table 10 questions your awareness of operations issues with regard to responsibilities. Table 11 forces you to benchmark subject by subject with what is considered best.

WHAT CAN BE EXPECTED OF OPERATIONS MANAGEMENT?
By D. Tijunelis

Table 10: What is going on?

Aspect of Operations Function	What is the Current Activity and Level of Attention?	Updated or Old?	Who Knows?	Action Needed
Operations Strategy				
Project Management				
Process Analysis & Job Design				
Product Design & Process Selection				
Quality Management				
Supply Chain Strategy				
Forecasting				
Resource Planning & Control				
Inventory Management				
Operations Scheduling				
Regulations: Safety / Environment				

MANAGING TECHNOLOGY
DEPENDENT OPERATIONS

Table 11: Benchmarking Against The Best

Pertinent Aspect of Operations:	How do we compare with the leaders?		
	By word of our employees or general opinion in the industry	By industry publications about competitors or our firm	By our suppliers, customers, and consultants
Operations Strategy			
Project Management			
Process Analysis & Job Design			
Product Design & Process Selection			
Quality Management			
Supply Chain Strategy			
Forecasting			
Resource Planning & Control			
Inventory Management			
Operations Scheduling			
Regulations: Safety / Environment			

WHAT CAN BE EXPECTED OF OPERATIONS MANAGEMENT?
By D. Tijunelis

References

1. Terry Hill *Manufacturing Strategy,* Irwin/McGraw-Hill, NY,
2. Charles W. Hill and Gareth R. Jones, *Strategic Management: An Integrated Approach*, 6th ed., Houghton Mifflin Co., Boston, 2004.
3. James P. Lewis, *Fundamentals of Project Management*, AMACOM, NY, 1997.
4. Norman Gaither and Greg Frazier, *Production and Operations Management*, 8th ed., International Thomson Publishing, 1999.
5. Jay Heizer and Berry Render, *Principles of Operations Management*, 3rd ed., Prentice Hall, Upper Saddle River, NJ, 1999
6. Richard B. Chase, Nicolas J. Aqilano, and F. Robert Jacobs, *Operations Management for Competitive Advantage*, 9th ed., McGraw-Hill, Irwin, Boston, 2001.
7. James R. Evans and William M. Lindsay, *The Management and Control of Quality*, South-Western, Cincinnati, 2002.
8. Jack Yurkiewicz, "Software Survey: Forecasting", *ORMS Today*, Vol. 30, #1, February 2003.
9. Eliyahu M. Goldratt and Jeff Cox, *The Goal*, Revised Ed., North River Press, Inc., NY, 1986
10. W.A.Burgess, *Recognition of Health Hazards in Industry*, 2nd Ed, John Wiley&Sons Inc., 1995

Suggested Additional Reading

- Rafael Aguayo, *Dr. Deming; An American Who Thought the Japanese About Quality*, Simon & Schuster, New york, 1990
- Matt Barney and Tom McCarty, *"The New Six Sigma"*, Prentice Hall PTR, Upper Saddle River, N.J., 2003.
- Jack R. Mredith and Scott M. Shafer, *Operations Magement for MBAs*, John Wiley & Sons, Inc., New York, 1999.
- Ravi Anupindi et al, *Managing Business Process Flows*, 2nd ed., Simon & Schuster Custom Publishing, Needham Heights,MA, 1998.
- Richard J. Schronberger and Edward M. Knod, Jr, *Operations Management: Customer-Focused Principles*, 6th ed., Irwin, Chicago, 1997.

MANAGING TECHNOLOGY
DEPENDENT OPERATIONS

- Lee J. Krajewski and Larry P. Ritzman. *Operations Management: Strategy and Analysis*, Addison-Wesley Publishing Co., 4th ed., Reading Mass.,1996.
- ACGIH (1992), *Industrial Ventilation, A Manual of Recommended Practice,* 21ed., Committee on Industrial Ventilation, American Conference of Governmental Industrial Hygienists, Cincinnati, OH.
- NIOSH (1978), *Pocket Guide to Chemical Hazards*, NIOSH Pub. No. 78-210, National Institute for Occupational Safety and Health, Cincinnati, OH.
- Roach, S.A. (1992), *Health Risks From Hazardous Substances at Work*, Pergamon Press, Oxford.
- AIHA (1993), *Who's Who in Industrial Hygiene*, American Industrial Hygiene Association, Fairfax, VA.
- Roger G. Schroeder, *Operations Management: Decision Making: Decision Making in the Operations Function*, 4th ed., McGraw-Hill,Inc., NY, 1993.
- Berry Render and Jay Heizer, *Principles of Operations Management*, 2nd ed., Prentice Hall, Upper Saddle River, NJ, 1997.
- Kiyoshi Suzaki, *The New Manufacturing Challenge*, The Free Press, New York, 1987.

Chapter 4, Lead-in:

LEADERSHIP, MOTIVATION AND TEAMWORK:
The "Software" of Organizational Success

This chapter addresses the handling of participative versus authoritarian management style in an operations environment. In doing so, invariably leadership and motivation become the key factors. Provided here are some basic organizational behavior management tips. They may be new to some and a reminder to others. Be sure to consider them just in case. When confronted with a challenge beyond ones knowledge, friends and cooperation are the best medicine. Teamwork is a must. A contingency model of leadership suggests considering what is your power of organizational position, your leader-member relationship, and the task definition. Having thoughtfully and objectively considered these, the choice of the best management style to use in a new environment may become obvious.

Management of teams have a "by-the-book" aspect which can be helpful, as well as examples that may be transposable to your situation. One of the most successful leaders, Jack Welch, the former CEO of GE, speaks of looking at business as a game. If it helped him, maybe it will help you. Examine your situation, compare, benchmark, look in the mirror.

In the automated, digital, and virtually integrated industrial environment today, the definition of operations implies a critical dependence on technology management. This means that every manager should be "manager of technology dependent operations." Therefore, the operative words are "technology dependent." It may mean your strength depends on others. Build relationships!

MANAGING TECHNOLOGY
DEPENDENT OPERATIONS

Chapter 4

LEADERSHIP, MOTIVATION AND TEAMWORK: THE "SOFTWARE" OF ORGANIZATIONAL SUCCESS

Matthew Puz

> *Strong leadership... culture...teams ... motivated people... represent the "software" of an organization.*

How many times throughout the annals of business history have companies that on paper should have enjoyed long periods of success have in practice failed to do so? For some reason, companies fortunate enough to have the best combination of resources, products, infrastructure, market share, talent and the like have all too commonly been trumped by competitors that come up with better products, service, and processes growing out of superior technology management. They cede their leadership positions to industry rivals that emerge as being better able to use technology to fulfill market demands. The advantage may not be reflected by financial measures or operations hardware. It may be in the building blocks of the "soul" of the organization. These companies have strong leadership, healthy corporate culture, high performance teams, and highly motivated people. The operational analogy could be to the "software" of an organization, which is what builds corporate

LEADERSHIP, MOTIVATION AND TEAMWORK
By Matthew Puz

culture and long-range success potential. It is the means by which technology-dependent operations (TDO's) strategies are implemented. Far too frequently, the concept is labeled as "touchy - feely" and only given secondary priority by the technical side of organizations, which are commonly more comfortable with tangible, physical reality.

Winning organizations have been able to maximize their performance and potential by capturing a competitive advantage through management systems that can adopt next generation technologies at increasingly rapid rates. Three formidable leadership challenges face modern day executives in managing TDO's. The first is to use what they have effectively. The second is to ensure that they actively pursue opportunities aimed at developing next-generation technologies as process improvements or marketable products or services. The third is to successfully drive a sufficient number of these opportunities towards fruition. All this being done at a frantic pace, while simultaneously managing a wide variety of ongoing initiatives. In this latter case, one example is to accomplish this through effective project management.

> *Leadership challenge:*
> *...use what is*
> *...see what can be*
> *...drive to fruition.*

LEADERSHIP "BY THE BOOK"

At an elementary level, management textbooks treat the topic of leadership as "Leadership and Supervision." More advanced treatments address it as "Leadership and Motivation." In an organization development context built on human behavior considerations, textbooks look at leadership as a process that has formal powers and informal traits to influence work outcomes. Leadership and motivation works in pairs, just as software cannot be written without a code. In a modern world, a leader has to have motivated followers to be an effective leader else they risk not tapping their followers' potential.

MANAGING TECHNOLOGY DEPENDENT OPERATIONS

> *...a leader has to have motivated followers to be a leader.*

So, what is leadership? Ask 100 people you can get 100 different answers with all of them being at least partially right. It is often said that leadership is a lot like pornography in that although difficult to define, you know it when you see it.

Defining Leadership: There are many books on the subject of leadership. Stogdill's *Handbook of Leadership* is 856 pages long (1). With so many leadership concepts in circulation, where does one begin to try to put their arms around so broad and difficult subject to grasp? The choice depends on your situation.

> *...leaders need to be good listeners...*

Effective leaders possess and exert an ability to positively influence and motivate those around them to act in ways that add velocity to their organizations in bringing about high levels of performance and corresponding achievement both at the organizational and individual levels. Through the effective exercising of their influence, such leaders are able to successfully impact the extent to which their organization's resources are harnessed, leveraged, and directed. This opens the door for an approach to lead a technical staff – specifically:

-Knowing the traits and capabilities of their followers and being flexible when necessary.
-Appreciating when and where leaders need to be good listeners and even followers themselves.
-Setting an example for others in establishing and retaining desirable protocols related to individual and interpersonal behaviors.
-Setting direction with regard to appropriate areas upon which the organization's resources should be focused.

LEADERSHIP, MOTIVATION AND TEAMWORK
By Matthew Puz

Leadership Styles: The seemingly limitless number of leadership models only rivals the number of leadership definitions that abound. As the general manager transitions into a technology leadership role, it is difficult and arguably unwise to simply advocate a particular leadership style. Just as singers must "find their sound" within the basic intonation and vocal framework that nature has blessed them with, each person must perform their own introspection to explore, discover, and enhance their particular leadership style.

> *...unwise to simply advocate a particular leadership style...*

1) **Either-Or Model:** A traditional model describes a leader's style as either *dictatorial* or *democratic* (authoritarian or participative). The implications should be obvious. Suffice to say that those practicing dictatorial approaches are more prone to falling into the trap of micromanagement, while those adopting democratic leadership styles have better prospects for cultivating more motivated and autonomous followers.

2) **A Continuum Model:** A more detailed model examines the continuum between dictatorial and democratic leadership - considers four style groupings. This approach, besides the extremes, recognizes the two sub-groups of *benevolent-dictator* and *consultive-democratic* leader styles. The former dictates from his or her point of view, and may not listen well or accept the followers views. The latter does not dictate but advises. The four-style model has been widely used by organizational behaviorists in industry for general organization development purposes.

3) **Contingency or Situational Model:** More sophisticated models account for the circumstances that the leader has to face. As such, these should fit TDO's better. One considers the choice of some form of authoritative vs. participative style on the basis of

MANAGING TECHNOLOGY
DEPENDENT OPERATIONS

- how defined is the objective,
- how much of formal authority the leader has, and
- what are the leader-followers personal relationships.

> *style on the basis of*
> *.... objective,*
> *.... formal authority,*
> *.... personal relationships*

As such this model is well suited for technology management situations. Military examples can easily be recognized. A respected platoon leader ordering a soldier to capture a machine gun nest may take one style; a young project leader investigating totally new product development opportunities in a strange country may adopt a different style. In its most recent versions it considers situational factors that take into account people's professionalism crossed with the task focus for a range of situations. This is the subject of the following discussions.

Leadership Style To Fit The Situation: A key underpinning of effective leadership relates to the need to adapt one's style to the particular situation. Pertaining to this, Paul Hersey and Ken Blanchard's landmark "Situational Leadership Model[®]" (2) is considered a tried and proven framework for enhancing the ability of leaders to effectively diagnose the situation and customize their leadership approach accordingly.

> *... "readiness" of followers...*
> *...are they able?*
> *...are they willing?*

In general, the model is based upon the presumption that leaders need to view the manner in which they lead people through the lens of the particular situation at hand. Prior to initiating a specific approach for leading others, the leader should assess the level of "readiness" of those who are to follow. Examine two dimensions – (1) The extent to which followers are "able" to accomplish a specific tasks and, (2) The extent to which they are "willing" or "confident" in their ability to accomplish the task (See Table1).

Table 1: Staff's Circumstances and Leader's Options

Circumstances are **GOOD**		Circumstances are **BAD**	
Staff:	Leader:	Staff:	Leader:
❑ Able <u>and</u> willing ❑ Motivated	❑ Confirms goals ❑ Delegates ❑ Tracks progress	❑ Unable <u>and</u> unwilling, ❑ Insecure not motivated	❑ Dictates, ❑ Seeks organizational options
Rely on <u>delegating</u>		*Use your <u>authority</u>*	
Circumstances are **MARGINAL**		Circumstances are **MARGINAL**	
Staff:	Leader:	Staff:	Leader:
❑ Able, <u>but</u> unwilling ❑ Insecure	❑ Understands their needs ❑ Works with them ❑ Motivates	❑ Unsure of ability, <u>but</u> willing and motivated	❑ Facilitates training ❑ Provides support ❑ Builds confidence
Focus on <u>motivating</u>		*Focus on <u>facilitating</u>*	

In applying this model effectively, leaders must tailor their style and approach to the appropriate stage that those they are leading fall along the respective "Able" and "Willing / Confident" continuums. Specifically, for persons demonstrating minimal ability to perform tasks, leaders need to provide a high degree of "task-related" support and guidance. Conversely, for persons demonstrating low degrees of willingness, confidence or motivation, leaders must provide a higher degree of support from a "relationship" standpoint.

> *...make an accurate assessment of the level of "readiness"...*

Failure to make an accurate assessment of the level of "readiness" their followers possess or underestimating the extent to which they may require supportive behavior may result in the leader overestimating a person's capabilities to effectively perform given tasks at hand. This model represents a tool capable of enhancing leadership effectiveness almost immediately by placing a balanced emphasis on the "people" aspect of management and "task emphasis" inherent in a technical environment. Review of this model will raise an awareness of the importance of assessing the respective "stages" of professional and personal development of followers.

The stages that people fall into along the main two dimensions that comprise the building blocks of this simple model vary across the full range. The dimensions of "readiness" can be described by subjective measures. The "Ability" for a given individual may include type and breadth of knowledge and exposure in a given area as well as the extent of "hands-on experience" and skills. A measure of "Willingness" and "Confidence" for an individual include how secure and motivated they are. It may also reflect their level of general interest in a given area, overall attitude and/or level of self-confidence. The latter case is impacted by both one's own personal make-up as well as by the nature of the organization's culture and its corresponding degree of tolerance for "failure." (See Table 2).

Table 2: Situational Leadership Issues

☐ Do you understand the technical challenges and skills required?
☐ Do you understand the technical capabilities of your staff?
☐ Is internal or external consulting help available to assist you?
☐ Is your staff confident in their ability to handle the tasks?
☐ Are the goals made clear to them?
☐ Are there training opportunities for the staff unsure in their abilities?
☐ Is your able staff willing to take on the tasks you are asking of them?
☐ Do the goals for the "willing" and "able" staff match yours?
☐ Do you relate well with the staff you have assigned to the tasks?
☐ Do you understand the reasons why the able staff is "unwilling"?
☐ Do you proactively look for opportunities to motivate?
☐ Do you have clear authority to formally provide support as needed?
☐ Do you have the option to reorganize?

Taken together, the placement of increased emphasis on the "people" aspect of management as part of an organization's "software" coupled with the benefits that the practice of a situational leadership model holds, represents a solid foundation upon which effective leadership can begin to be built.

LEADERSHIP IN PRACTICE

What Really Makes a Leader?

What exactly is it that distinguishes and differentiates companies that seem to consistently develop or adopt viable new technologies more quickly and more effectively than their competition? Senior management often says that their organization is successful because "People are our most valuable asset". Let's be honest. As most of us hear these high sounding words, it is difficult

> *... easy to fall prey to the notion that "others are like me"*

MANAGING TECHNOLOGY
DEPENDENT OPERATIONS

to not feel at least a slight tint of "Sure. We've heard that one before". Lip service. As we drill down looking for organizations that are proficient in the development or application of new technology streams, one common trait seems to emerge within which a measure of hope resonates. Leading companies, in large part, truly do seem to practice leadership based on a "people first" approach to management. In short, these are people-driven organizations. Effective leadership practices demonstrated by such organizations represent a critical component in the process of developing and implementing successful technologies.

> *Not everyone wants to be a leader.*

To what extent do organizations instinctively structure and manage themselves in a fashion that fosters high performance in their people? Herein largely lies the responsibilities of general management to maximize the levels of contribution of all people within the organization – In short, to promote good leadership.

However, not everyone wants to be a leader. Although on the surface being a leader sounds inviting, prompting the common response, "Sure, why not?", in actuality leadership effectiveness comes at a price. It hinges upon the level of commitment and investment that one has and is willing to make in ensuring that their organization's human capital is fully tapped into and leveraged.

The concept of leadership may be interpreted differently by the technical staff than by general management. Re-visiting the situational model in the context of human nature, in assessing the level of a "readiness," it is all too easy to fall prey to the notion that "others are like me." This common pitfall makes the effective adoption and practice of the situational model previously discussed more difficult.

At the individual level, acquiring an accurate understanding of one's own particular style represents a major goal. The discussion to follow is

LEADERSHIP, MOTIVATION AND TEAMWORK
By Matthew Puz

partially intended to prompt a healthy introspective process that sheds light and insights into one's own interpersonal style and the impact it has on one's leadership effectiveness.

Bringing out the Best in People: In many ways, the challenges facing leaders of technology development today are similar to those of agents representing creative artists of all types. For TDO's, success is deeply rooted in their ability to tap into the creativity of those involved with innovation and implementation of promising technology-based initiatives. Done properly, people become instilled with self-confidence borne out of the trust demonstrated by their superiors. Such individuals feel that limitations are lifted on their potential to create, excel, and contribute above and beyond levels they formerly thought to be possible.

The bottom line? A big part of the role of leaders is to continually look for the winners within their staff and to tap their full depth and breadth of potential. People that can say that they are working to bring out the best in their staff as a regular part of their daily activities have taken an important step along the road towards effective leadership. Recognizing that success often depends upon tapping areas beyond one's own expertise, taking a broad view of both the responsibility and opportunity to bring out the best in others is key.

Creating a Climate of Unselfishness: As current and would-be leaders attempt to better understand their leadership styles, one characteristic emerges as being central to the practice of effective leadership: the need for leaders to demonstrate "unselfishness" at both the personal and professional level towards those they are attempting to lead.

> *...you depend on initiatives in areas beyond your own expertise.*

MANAGING TECHNOLOGY
DEPENDENT OPERATIONS

This specifically pertains to the day-to-day priority decisions that people in operations have to make. Often times attention is given disproportionately to staff whose names are associated with high visibility issues and accomplishments. As a result, this reinforces people's tendency to devote the bulk of their efforts towards the accomplishment of measurable and tangible initiatives. Once preoccupied with the state of progress associated with one's own "deliverables," it is difficult for people to simultaneously serve as a vehicle geared towards tapping into and bringing out superior levels of performance in others.

The unwritten rules of society serve to reinforce the creation of behaviors that are more heavily inwardly focused on individual achievement as opposed to outwardly focused towards assisting others. From an organizational standpoint, a working environment that promotes and positively reinforces this type of behavior creates an "every man for himself" based culture. Within such an environment, the creation of high performance teams becomes a difficult task.

> *...transition from "Player" to "Coach."*

Naturally, especially within technology dependent operations, fostering initiative directed towards tangible achievement is highly desirable. However, doing so requires current and emerging leaders to strike a delicate balance between "doing" and "leading" as they transition from "Player" to "Coach." For individuals to be successful as leaders, they must ensure that the amount of time and effort they devote in their daily activities towards nurturing professional development and performance in others is at least at a minimum threshold level.

> *...the concept of "Servant- Leader."*

Amidst the increased competitive conditions, the need for organizations to be able to "do more with less" should prompt foresighted individuals to strive to cultivate

LEADERSHIP, MOTIVATION AND TEAMWORK
By Matthew Puz

situations where more of their work is accomplished through the efforts of others. Relying on one's own efforts and activities alone in this day and age will not work.

Servant Leadership: Taking personal and professional unselfishness a step further, the concept of Servant Leadership is based upon a seemingly counterintuitive notion. Instead of employees holding subordinate positions and serving their superiors, an organization's "higher ups" hold a primary responsibility to serve those *below* them. Part of the magic in overcoming the seemingly natural reluctance that people have towards putting themselves in the service of others seems to lie in getting started down the road towards this practice. The first unselfish act of good faith where the leader asks, "How can I help?" holds tremendous potential to establish a powerful new relationship and protocol.

Leadership lessons from Mount Everest: A particularly striking case in point testifying to the power of servant leadership occurred on May 24, 2001, when a man by the name of Erik Weihenmayer climbed to the top of the world by scaling Mount Everest.

> *...behind the scenes, ... "case of servant leadership represented the driving force"...*

What makes this accomplishment even more unique, if not miraculous, is the fact that Erik is blind (3). Quietly, behind the scenes, a powerful case of servant leadership represented the driving force behind this incredible feat. Unselfishly, a man by the name of Pasquale Scaturro who himself had already scaled Mount Everest, came upon the idea of putting together a climbing team that would focus on giving Weihenmayer this opportunity of a lifetime. *"Hey dude, you ever climb Everest? Do you wanna?"* And so it started. Two years later, a sightless Erik Weihenmayer stood atop the highest mountain peak in the world.

MANAGING TECHNOLOGY DEPENDENT OPERATIONS

For Weihenmayer, the cover of "Time" magazine and numerous other accolades followed. Scaturro and the rest of his team received some footnotes but not much more. So what makes a person like Scaturro initiate, undertake and devote himself to such a major endeavor? Especially when he knows full well that the dedication, his time, the expending of both physical and emotional energy, and the literal risking of his life will very likely never result in receiving accolades anywhere near to those bestowed upon the recipient of his "unselfishness?" Clearly, a major part of the drive to champion such a cause is based purely upon the satisfaction derived from having brought out the best in someone else. Furthermore, it relies on the corresponding belief that laying oneself on the line for someone deserving of the opportunity to "shine" is the right thing to do.

> *... ripple effects of technical innovation...*

And what about the ripple effects? How many blind people, let alone deaf, learning deficient, disabled, or those "challenged" in any of a number of different ways, now hold themselves up to a higher standard in terms of what they believe themselves to be capable of achieving? And when they perform their own "gut checks" and inch the bar up at which they set their own personal standards just a little bit higher, the person staring back at them in the mirror may very well be Erik Weihenmayer. But standing behind Weihenmayer, out of sight, is the person who created Weihenmayer's opportunity to scale the heights that he himself never imagined possible - Pasquale Scaturro and the rest of his climbing team. Knowing this, both Scaturro and his team, in their own un-glamorized and under-publicized way, must find a tremendous sense of satisfaction for unselfishly having put themselves in the service of Weihenmayer. Likewise, for business leaders embracing the practice of servant leadership in dedicating themselves to elevating the performance of others, they too can find great satisfaction from contributing to and witnessing the professional development of those under their watch.

LEADERSHIP, MOTIVATION AND TEAMWORK
By Matthew Puz

Unselfish leadership can have positive ripple effects in enhancing technical innovation initiatives throughout operations. For tech leaders operating within TDO's, embracing this practice can be difficult. Operating within environments that place high value on being precise and correct with little margin for error, impatience and intolerance can be either consciously or unwittingly encouraged. As such, making a conscious effort to make "serving others" a priority is needed. When your gut causes you to want to "go for the throat" on a greenhorn whose project has screaming deficiencies, stop, count to 10, then ask "How can I help? Let's plug these holes." Try it. Give until it hurts. Due to the importance of your efforts being genuine, start with somebody whose professional development you deem important to you to generate some momentum as a springboard for potential wide spread application of the practice.

High-Touch Leadership: In pursuing leadership effectiveness, recognizing the important role that leader-member relationships have as they pertain to having contact with people is critical. In some respects, it simply boils down to generating as many "leadership opportunities" as possible to exert one's positive influence over others.

> *...hear unfiltered information ....to feel the "pulse...*

Studies have revealed that effective leaders devote an extremely high percentage of their time to creating interactions with people as opposed to "working alone." Furthermore, the most effective leaders exhibit a tendency to cultivate networks that extend beyond persons that are encompassed by the boundaries specified by their organization's formal structure once removed from their "home areas" - perhaps their technical staff.

One emerging practice geared at creating "leadership opportunities" being used more commonly is the holding of monthly meetings by senior level executives with employees having birthdays falling in a given month. These "birthday meetings" encompass virtually all areas and management levels within the organization. The intent is to create a forum within which employees have access to senior executives. Within such an environment,

MANAGING TECHNOLOGY
DEPENDENT OPERATIONS

> *What's your... "People-Time Quotient"?*

senior level executives have the opportunity to hear unfiltered information and feedback first hand from employees, thus enabling them to feel the "pulse" of the organization at multiple levels. Additionally, they are afforded the opportunity to promote their agenda, recognize accomplishments, and communicate new management direction and initiatives in a "face-to-face" fashion to a broad cross-section of employees - in short, to lead and positively exert their influence. A further benefit of this format allows employees to hear things "from the horses mouth." This forum for communication contrasts sharply with the more sterile and impersonal written electronic (e-mail) or grapevine communication modes. Through this process, top management can effectively demonstrate their commitment to a "high touch" leadership approach throughout their organizations. Again, the benefits are not free as they demand that a commitment be made to prioritizing and practicing a "high touch" approach to leadership.

Time Tracking: "Of coarse I know where my time goes. Doesn't everybody?" There is some strong evidence that suggests that in spite of the insistence on the part of managers that they are acutely aware of how their time is allocated, they often have a distorted, inaccurate perception as to how their time is actually spent. Consider contributing factors of Table 3:

Table 3: Origins of Leader's Time Allocation

1) *An almost imperceptible "drifting," loss of control, over how one's time is spent due to the demands other are making upon them.*
2) *A disconcerting sense that one is being drawn into activities that are inconsistent with a leadership role.*
3) *The existence of a "blind spot" to fall prey to task-specific activities that should be delegated to others more suited to perform them.*

The combined impact of these instances and circumstances often lead to an inappropriate allocation of a manager's time along the continuum of "leading" versus "doing." The real danger of this resides in instances in which leaders are inappropriately drawn into the actual performance of task-specific activities. This robs them of the time they have available to devote to the more leadership-based activities of monitoring, supporting, and coordinating the activities of others under their watch. This dynamic, especially common with people of a technical background, results in the limiting of leadership opportunities. Further, it impedes a leader's ability to maximize the number of high quality, "high touch" interactions they have with their people.

> *Note when you are particularly susceptible to time allocation "drift" ...*

How do you spend your time? Check yourself. Try some form of time tracking on a periodic basis. Note when you are particularly susceptible to time allocation "drift" from leadership-based activities and becoming too heavily focused on the actual performance of task-specific activities on an individual basis. For the lack of a better term, call it your *"People-Time Quotient"* - the amount of time you are spending with others under your watch in mentoring and directing their respective activities.

By way of example, a simple yet effective time tracking form is provided in the Appendix. Some form of it, adapted for your situation, will provide valuable insights how your time is being allocated.

Making the "Doer" to "Leader" Transition: Emerging leaders within technical organizations face making the transition from active practitioner of technology over to a leader of technologists. In essence, this means making the transition from a "doer" to "leader". Although this particular challenge is common amongst a variety of business disciplines, it often represents a particularly vexing problem for technologists.

MANAGING TECHNOLOGY DEPENDENT OPERATIONS

> *...great technologists do not necessarily make great managers and leaders...*

Within operations that depend on technology, people that have stood out as being superior performers excel by virtue of their outstanding performance in a narrow area of technology. For that it took special training and skill. They commonly work their way up into management positions through their technical achievements. The importance of these technical contributions often lead superiors to "reward" them by increasing their level of responsibility. In doing so, a great deal of responsibility thus falls upon these superiors to ensure that those whom they promote possess the proper perspectives as to how their new role and its corresponding responsibilities have changed from their past historic "technology-practitioner" role. "Real" technologists do not necessarily make great managers and leaders. Of course, this is certainly not restricted to the case of the technologists. Take the case of the superior salesperson that is ill equipped to handle the role of a sales manager. For persons transitioning into management positions, it is important to recognize that the combination of experiences, traits, and tendencies that have potentially served them well throughout their careers may very well run counter to the skill sets they now require to become effective leaders and managers. (Figure 1)

Implications for Technologists in Transition: So what are the lessons for leaders within technology-dependent organizations? Like a carpenter equipped only with a hammer, irrespective of the job at hand, the tendency is to reach for the hammer. Likewise, technologists transitioning into leadership positions often find it difficult to resist the lure of being drawn into the details of specific research, development, or engineering projects. Certainly the unique experiences and knowledge that technology leaders have amassed in their careers need to be tapped into and leveraged for the betterment of the organization and its people. The key, however, is to strike a proper balance between transferring knowledge to others to enhance their effectiveness while simultaneously guarding against stifling their initiative by becoming too "hands on" and involved in the details. It is important that

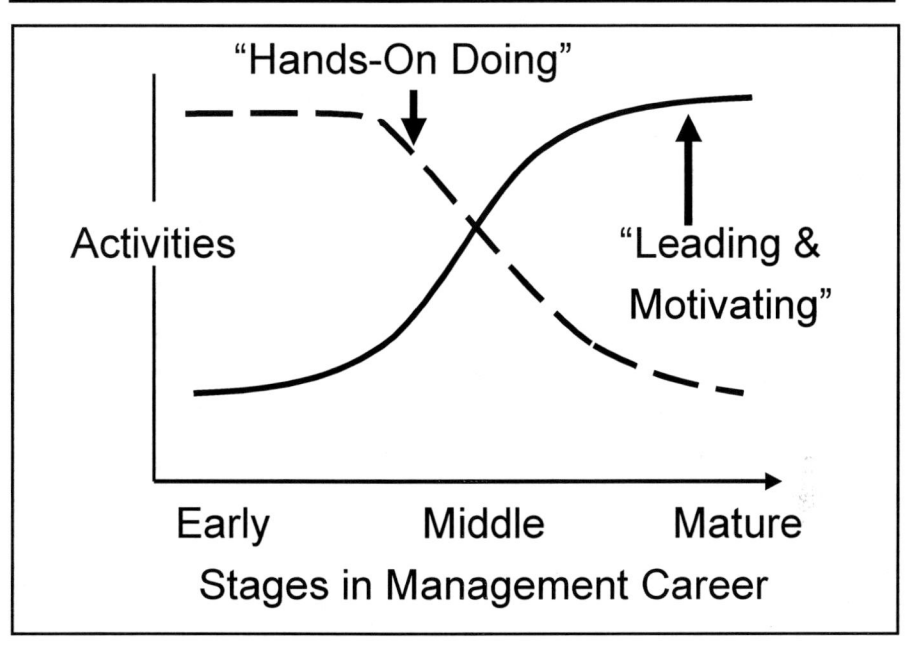

Figure 1: The "Player" to "Coach" Transition

technologists in leadership roles recognize that they are on a "slippery slope" that can easily cause them to fall prey to their natural instincts by drawing them too far into the "doing" side of the continuum as opposed to the "leading" side. By clinging too heavily to a "hands on" approach, important lost opportunities and undesirable consequences often occur:

1) The sending of undesirable signals to those under their watch to first enlist their leader's full capabilities as a first option prior to when the leader's involvement is clearly warranted.

2) The establishment of a protocol that impedes and undermines the individual leader's development wasting time that could be spent "leading" a broader range of people.

> *...guard against becoming too immersed in the actual "hands on" activities....*

MANAGING TECHNOLOGY DEPENDENT OPERATIONS

An inability to break free of the "hands on" approach to management is a "lose-lose" proposition for all involved.

Giving People a Chance to Rise Up: In nature, when a void is created, the respective constituencies surrounding the affected area exhibit a tendency to fill that void. The translation of this being, "Nature abhors a vacuum." So how does this concept relate to leaders of technology management and to those that are transitioning into leadership positions?

Mort Meyerson's early indoctrination into the EDS culture all the way back in 1967 holds some interesting insights. In spite of his lack of experience at the time, Meyerson was thrust into the role of developing a major proposal for Blue Shield to process their Medicaid claims. Despite his apprehensions and uneasiness, Meyerson took comfort in the knowledge that the actual "pitching" of this first presentation to the client would be done by EDS founder himself, Ross Perot – the ultimate salesman.

> *...lesson to be taken - people do adapt.*
> *... may lead to unusual leadership skills.*

Shockingly, a mere 1 hour before the scheduled presentation was to take place, Meyerson discovered that Perot was (very) unexpectedly "out of town" leaving him no choice but to make the pitch himself. Perhaps Meyerson was being punished for something? Nope, that's not it. Although Ross Perot was undisputedly the best equipped to make the presentation, had he done so he would have missed out on a golden opportunity to create an invaluable professional development experience for Meyerson. Specifically, his absence created a "vacuum" forcing Meyerson to step up and raise his performance level. By simply not "showing up," Perot orchestrated an important leadership opportunity: "Sink or swim, welcome to EDS!" Meyerson swam and went on to become its president. (4)

The lesson here is that tech leaders should first consider whether their "hands on" involvement is required before jumping in and rolling up their sleeves. Specifically, will doing so cause you to miss out on professional development opportunities for others – one that will allow them to "rise up," elevate their performance levels, and experience increased confidence? If so, make yourself scarce and watch carefully from a distance.

As it relates to your own situation, consider the following Personal Introspection questions below.

Table 4: Giving People a Chance to Rise Up
- Personal Introspection

1) Do you have faith in your subordinates to rise up and fill any technology vacuums you may create?

2) How is your time being allocated along the continuum of "doing" versus "leading" technology dependent activities?

3) Given your position, does this allocation of your time seem appropriate?

4) Are you too heavily involved in "hands-on" activities? If so, to what do you attribute this? A) Wanting to put your "fingerprints" on tangible TDO-based results; B) Failure to foster a protocol of autonomy with your people; C) Allowing subordinates "manage up" activities that are more suited to them.

5) To what extent are you "creating vacuums" for others to rise up and fill thus fostering participation and professional development for them?

6) To what extent are you considering the particular situations at hand facing your staff and tailoring your approach accordingly?

In summary, as technology leaders wrestle with the questions above, inevitably they must answer the question as to how best each situation they encounter should be handled. Their goals should be to establish desirable protocols that lead towards professional development of their subordinates, allow for an appropriate allocation of their time, and guard against the compromising of task accomplishment.

MOTIVATION "BY THE BOOK"

If the leader is to lead, there needs to be motivated followers So what are the motivational tools that a leader can use to maximize the prospects that followers will, in fact, follow and respond positively to leadership influences?

> *There is a "jungle" of motivational gurus...*

There is a "jungle" of motivational gurus and models. Some find solid ground in behavioral theories, while others last as long as the charisma of the advocate is sustained in the public's eye. The more popular management textbooks speak of content and process theories. Content theories try to explain why do people do what they do while process theories identify how motivated behavior is initiated, sustained, redirected, or terminated.

> *You better be sure you have the right person for the job.*

Only a few theories stand out for practical applications in the management of technology dependent operations. One is an old content theory made famous by Maslow that argues that people are motivated in stages along a progressive hierarchy of needs (5). These start with physiological needs to live, then general security, followed by social recognition, esteem, and ultimately self-actualization. Obviously, at the lowest physiological stage one is unlikely to motivate people with an opportunity to be creative when their focus is on getting something to eat and finding shelter. Near the top of the needs sequence, when many basic needs are met, recognition of achievement may be the prime motivator. A pat on the back, an opportunity to publish, or an award in front of peers may be what is needed in a technology-dependent organization. On the other hand, at the top stage, where all other needs are met, even recognition may not be effective. Here you may have to tap into an individual's self-fulfillment or their love of the work as the prime motivator. Motivation is anything but a "one size fits all" proposition. As such, you better be sure you have the right person for the job. Table 5 raises some key issues.

Table 5: Needs-Based Motivation

1. **Does the pay meet the basic needs of the technical staff?** Yes__, No__, ?__.
2. **Are jobs considered safe and positions stable, secure?** Yes__, No__, ?__.
3. **Is there competition or conflict among peers?** Yes__, No__, ?__.
4. **Do technical achievements get adequate recognition?** Yes__, No__, ?__.
5. **Are well-rewarded specialists in right assignments?** Yes__, No__, ?__.

One of the more noteworthy and applicable process theories considers the expectations of people at work – termed *Expectancy Theory* (6). It suggests that that motivation is a product of the following three basic factors:
1. The probability of one's efforts fulfilling the needed task requirements.
2. The value and desirability of the reward that achieving the objective will bring.
3. The perceived likelihood that meeting the requirement will actually bring the reward.

MANAGING TECHNOLOGY
DEPENDENT OPERATIONS

Carrying this forward, the resultant motivation is the product of the three factors:

| Motivation = Success Probability x Reward Desirability x Reward Likelihood |

Interestingly, the relationship between the three factors is multiplicative. That is, if the value of any one of these factors goes to zero, then the individual's motivation will be zero. In short, all of these factors are of crucial importance thus allowing no leeway for any of them to be ignored.

For operations management, this theory can have direct application to the management of technologists. Consider it in tandem with the hierarchy of needs. As in the case of situational leadership, by considering what the specific key motivators, needs, and expectations are for each of those under your watch, it is possible to effectively tailor one's approach to motivation to the situation at hand.

Table 6: Motivating Expectations

1. Does your technical staff feel the need for the goal?   Yes___, No___, ?___.
2. Are the level of want for rewards viewed as being desirable? Yes___, No ___.
3. Are they confident that they can meet the given goal?   Yes___, No___, ?___.
4. Are their levels of their competency sufficient?   Yes___, No___.
5. Are they aware of requirements beyond their control?   Yes___, No___, ?___.
6. How do they perceive the level of risk?   Hi___, Low___.
7. Does your staff believe that administration of rewards will occur? Yes___, No___

Take a situation where a new assignment is made to a senior technical specialist who has received many awards in his specialty, has been well paid, and is near retirement. One can imagine that such a person has reached the stage of self-fulfillment. By the model a high level of motivation for an unfamiliar assignment may be low possibly because of lack of interest, but probably not because of the risk (or fear of failure). On the other hand, a junior technical staff member would be much more likely to desire such an assignment on the basis of pay or recognition potential, but may lack the confidence in their capabilities to successfully fulfill the task. Again, accurately sizing up the situation from a motivation and skills/capabilities standpoint is the first order of business for the tech leader.

Impact of Corporate Culture: In considering the discussion of the leadership and motivational concepts covered thus far, it is important to recognize the impact that the overall working environment existing within the organization has on these factors – that being an organization's corporate culture.

Corporate culture is often defined as being the collective set of observable and predictable behavioral patterns occurring within and practiced by an organization's employees. At its best, culture drives the organization and its operations towards desired behaviors that are understood and reinforced. At its worst, an unhealthy culture has potential to undermine an organization's effectiveness at both the technology-development and corporate levels.

As it relates to technical staff, a culture's accepted and reinforced behaviors are especially important as they represent the framework within which technology management takes place. It is well documented that corporate cultures evolve over extended periods of time and are perpetuated as veteran members encourage new members to adopt and practice the legacy of behaviors, attitudes, values, and norms that characterize their organization in their own behaviors.

> *...harness the collective efforts...*

Most individuals within large organizations cannot reasonably be expected to have a dramatic impact on its corporate culture. That is unless they are part of a senior management team. Just the same, for most it is still

MANAGING TECHNOLOGY DEPENDENT OPERATIONS

important to be cognizant of the impact they can have in influencing culture through their direct interactions with others. Like voters who often adopt the "Because I only have one vote, I don't have the power to make a meaningful difference" mindset, people too often lose sight of the true power that they do, in fact, possess. As in an election, although each individual only has one vote, all have the opportunity to *influence* the voting behavior of others , especially through the use of enhanced leadership practices.

Carrying this analogy over to organizations, people who choose to do so can, in fact, effectively exert significant influence over those they interact and come in contact with in fostering meaningful, albeit possibly small, changes in the behaviors, attitudes, and patterns of interactions amongst them. In doing so, they can play key roles in the creation of pockets of healthy sub-cultures, even within organizations that may be "culturally challenged." Tech leaders who find themselves working within such environments should not lose heart but instead should view themselves as representing the first domino along the path towards their organizations more fully tapping into its "software."

Team Building: "By The Book"

Effective teams are a crucial part of an organization's software making it imperative for Tech Leaders to be aware of the accepted techniques found in management textbooks and their real world limitations. Some of the published advice assumes that you can pick an "all-star" staff thus creating an ideal team. In most cases, it is simply not realistic to "start over" with entirely new work force. You may have to make do with what you have. This applies to all organizations whether they are large or small. To compete on any level, management needs to practice and rely on effective leadership to adapt, motivate, and pool their resources into focused teams.

In the climate of fast changing technology, "speed" represents the new cornerstone of competitive advantage and has become the key differentiator between industry "winners" and "losers." The ability of organizations to quickly configure and effectively manage the skill and knowledge possessed by their people has become a crucial element in building a road map to future success. By being proficient in this key area, TDO's can maximize their effectiveness and chances at being "first to market", "first to patent," "first to apply to business operations," etc. In doing so, they can minimize the risks of becoming "victims" of a changing technological landscape.

What Is a Team? A group of people having a high degree of interaction and seemingly behaving in a similar fashion related to what may look to be the same goals is not necessarily a team. The distinction gets at the root of what separates groups such as cliques intermingling social interests from efficient groups with work objectives. Teamwork is considered as either a group of workers led by a strong leader or as "work teams." The latter case becomes most effective when self-directed in relying on shared leadership. That is leadership that is unselfish and incorporates a learning process.

Although success in TDO's is greatly dependent upon the ability to build, manage, and motivate high performing teams, doing so may not always be easy. Tech leaders may have to navigate around limitations or deficiencies in key, but essential technical areas. Also, problematic interpersonal dynamics within teams or by specific individuals can undermine team success. Let's face it, people don't always play well together. Recognizing these limitations and the reality that team effectiveness won't simply happen on its own, tech leaders need to make a concerted effort to stack the deck in the favor of making it happen.

Action Item: Start by assessing the status of the teams you currently have in place or are about to form using the tools provided in the tables that follow.

MANAGING TECHNOLOGY
DEPENDENT OPERATIONS

Table 7: Survey for the Effectiveness of Teams in Operations

FEATURE	STATUS			REMEDY/ COMMENT
	YES	?	NO	
Has Relevant Skills				
Has Unified Commitment				
Communicates Effectively				
Practices Mutual Trust				
Has Effective Leadership				
Gets Adequate External Support				
Can Negotiate As Needed				
Sticks to Clear Goals				
Has a track record of success				

...you may have to do with what you have...

The Team-Building Process: Operations are managed as simply ongoing routine work or as challenging projects. The projects may stem from productivity problems, quality issues, a need for operating system innovation, or some response to customer complaints. While each has some peculiarities, a general road map of progressive steps should be followed. These typically are problem identification, problem analysis, cross-functional review for potential solution generation, assessment of pros and cons associated with each option, final solution recommendation, review and decision to act and implement. A team is likely to be formed to handle all or just some if these specific steps.

Question: Where is your challenge as it pertains to your current standing teams?

Table 8: Team Status or Needs

"Stage" - Operations Problem To Tackle :	Done Yes	Done No	To Do Individual Assignment	To Do Team In Place	To Do Team To Form
Problem Defined					
Problem Analyzed					
Solution Brainstorming					
Solution Identification					
Action Recommended					
Decision Confirmed & Implemented					

Stages of Team Development:
There are certain tried and proven methods that should be involved in any team building effort. By way of definition, a team can be described as "A small group of people with complementary skills who are committed to a common purpose, set of performance goals, and approach for which they hold themselves mutually accountable" (7).

...people with complementary skills, common purpose, set of goals ...mutually accountable...

In considering these team building blocks, familiarity with the stages that teams generally transition through is important. A widely accepted, classic model consists of the following five distinct stages (8):

Forming → Storming → Norming → Performing → Adjourning

In the absence of non-traditional team-building techniques, most groups commonly transition in some way, shape, or form through these stages. A general overview of these stages along with the practical management issues at each stage follows in Table 9.

MANAGING TECHNOLOGY
DEPENDENT OPERATIONS

Question: What stages are the teams under your guidance at? As you consider the discussion to follow for each stage, try to glean insights about "problem areas" you may be witnessing within each team. Note: A team can be in more than one stage at a time.

Table 9: Stages of Progressive Team Development

Team Stage	Where is your team?	Stage Description and Overview
Forming	_____ %	Team formation, member selection
Storming	_____ %	Struggle for control, power, roles & influence
Norming	_____ %	Team-based protocols established, more subtle undertones of Storming stage still lingering
Performing	_____ %	Team finds its "groove" and reaches its peak performance as a function of its effectiveness in instituting high performance team measures
Adjourning	_____ %	Upon goal completion, team's human resources re-deployment plans in place and initiated

Forming

The *Forming* stage is simply what the word implies – the initial gathering and formation of the working group. In this stage, candidate members either voluntarily express interest in group-membership or are "volunteered" by others, typically superiors. From an individual standpoint, team candidates often "leap ahead" mentally in considering what contributions they view themselves as potentially being capable of bringing to the team.

...rapidly bridge any technological gaps...

In doing so, they compare these to group expectations while also assessing the cultural dynamics and behavioral norms most likely to be present in the group. In short, they should answer the question, "Do we appear to have the makings of a good team as it pertains to skill bases needed versus available and the impact of individual member styles." Recognizing the importance of these factors, solid judgment about team member selection is a key factor in this equation.

Whatever the definition, in practice as a manager and a tech leader, you could have a number of options at your disposal to support effective team formation. At this stage you can choose to assign team members, unofficially encourage someone to join a team, or simply call for volunteers. On the other hand you may discourage someone from joining a team because of skill or interpersonal dynamics concerns. You may even choose to disband a team under certain conditions. Such decisions depend on the purpose of the team, its current standing as well as the potential contribution of the team to the organization, and likewise, by the individual to their respective teams. The real issue is whether you know the contribution potential of the individuals within the particular team setting you are forming and if you feel they can contribute meaningfully to it. As a tech leader responsible for the formation of teams, consider the questions of Table 10:

Table 10: Questions For Team Membership At The Start

Team:_____;   Team Member:_____

EXPECTED PARTICIPATION	Fit			How Critical?		
	YES	?	NO	Key	Med.	Low
1. Does the intended member understand the expected team performance strategy and have the necessary commitment towards it?						
Will commit to specific technical goals?						
Will accept a team approach?						
Will see the strategic value of team effort?						
2. Does the intended member acknowledge accountability?						
Will accept & appreciate mutual accountability?						
Will accept individual accountability - Not hide?						
Will not fall into "group-think" - Will speak up?						
3. Will the intended member bring synergy to the team?						
Will bring added, needed, technical skills?						
Will enhance positive problem solving skills?						
Will foster positive team dynamics - Not be a source of conflict in the team?						

Needless to say, technology focused teams should have a technical expert in the key role. However, in spite of their importance and key role, tech leaders should be aware that such people can become overly dominant in a team setting unless balanced by someone capable of facilitating a team approach. Technology-dependent operations are especially vulnerable to this common occurrence.

Storming

During the *Storming* stage, considerable "jockeying for position" typically occurs as emotions run high, often leading to tension and group member friction. These include the onset of conflicts over group leadership and authority and the potential formation of sub-coalitions within the group. Amidst Storming, questions related to contributions (and ability to contribute) often create further turbulence. Eventually, as tensions subside and group member expectations and roles are formed, members begin focusing on group goals despite these potential impediments to group success.

Although no one is likely to openly call attention to these potentially destructive interpersonal dynamics and may even consciously seek to conceal them, tech leaders need to be on the look out for them. The key is knowing when to step in to guard against a protocol of negative dynamics and inappropriate control and influence from occurring as opposed to when to allow the team to wrestle their way through Storming on their own. In certain cases, having brutally honest dialogue with individuals or clique coalitions forming within a team may be necessary. The message is simple – "The team comes first and destructive practices at the individual or clique level won't be tolerated. Get with the program or you're out." Often at the root of such problems lie abuse in one of the five basic power sources cited in Table 11. Used improperly, these power types can undermine effective team dynamics and corresponding performance.

Question: Are the sources of power in your team situations being used appropriately by team members and in the right proportions for the team to function effectively? To identify festering problems in your teams, consider their situations in light of Table 11.

Table 11: Power Factors Within A Team

Power Source	Description	Who has it?	Level Hi	Level OK	Level Low	Action Needed
Reward	Can offer tangible rewards to others					
Coercive	Can hint at or administer reprisals or punishment					
Legitimate	Appointed or elected role					
Referent	Identification with others in power					
Expert	Contribution & Control of unique knowledge					

Obviously technical experts are important in technology dependent organizations. However, in instances in which there are more than one "tech expert" on the same team, the "Not Invented Here" (NIH) Syndrome can rear its ugly head. Again, tech leaders need to be aware of when to step in and possibly having to resort to the step of communicating the need for people to "check their egos at the door" in the best interests of the team. The same, too, applies to the tech leaders themselves. Recalling the importance of giving people the chance to raise their performance levels, tech leaders need to be sure that the legitimate and potentially coercive power of their voices does not subvert the critical voices of the technical expert(s).

Hint: Consider missing an occasional meeting or deferring to a subordinate for their suggestions (in spite of your potentially having a strong opinion for a coarse of action) to create a "vacuum" for others to step up and fill and to enhance the freedom of expression within your teams.

MANAGING TECHNOLOGY
DEPENDENT OPERATIONS

The Functional Allegiance Dilemma: Historically, the allegiances by individuals to their home functional areas has often led to internal politics. The all too common end result of this has been the creation of disharmony between core functional areas and a general deterioration in the interpersonal organizational dynamics, the hoarding of information and the withholding of key resources required by other functional areas. A hypothetical example goes something like this: Sales and Marketing wants a "do all" product for the marketplace that Engineering is unwilling to take responsibility for the design and/or performance and/or for which Manufacturing is unwilling to commit to being able to reliably produce. The result is commonly an epidemic of "throwing the project over the wall" to sister functional areas. All this occurs despite the almost certain knowledge that the "rejection cycle" will cause it to only come back again. It is important that tech leaders not get caught off guard by the potential impact of functional area allegiances undermining team performance. The tools and techniques to follow in this chapter should go a long way to providing an effective arsenal to combat this, as well as other, challenges to team effectiveness.

Norming

In the *Norming* stage, a sense of balance within the group is established along with the emergence of interpersonal protocols. A measure of "social harmony" also commonly forms within the group. Having survived the strains and unpleasantness of the Forming and Storming stages, the team members often feel strongly inclined towards maintaining this new harmony often compromising the need for open and candid communication. A common pitfall growing out of this dynamic is the onset of "groupthink." This is characterized by a loss of a team's evaluative capabilities (9). Affected persons become hesitant to criticize other's ideas in the interests of maintaining group harmony. The common result is the watering down of healthy debate and dialogue thus undermining team effectiveness. Under such conditions within the Norming stage, although the team appears to function well in a social sense, it has yet to become high performing.

> *...the onset of "groupthink"- hesitant to criticize...*

Table 12: Technology "Groupthink" During "Norming"

Symptom	Yes	No	Possible Action
Feeling that the team is correct without question			
Belief in the team members unselfish motives			
Rationalizing not to disagree			
Sharing stereotypes – "engineers must know"			
Self-censorship – not to be negative			
Pressure from superiors for conformity			
Preconceived ideas – Limited options & opinions			
Illusion that all team members are in agreement			

Performing

Mercifully, as groups enter the *Performing* stage, its members emerge as a well-functioning unit. Having matured, the group is largely motivated by group goals and becomes strongly committed to building upon these strides despite any lingering or periodic relational strains that may exist or arise.

As previously established, "speed" has become the new differentiator in the 21st Century. For TDO's, we're talking about speed in areas such as product development, time to market, and implementation time for introducing new processes. To this end, assembling the right people with the right skills and knowledge together at the right time and properly managing and motivating them towards the achievement of key initiatives are central to continued success – in short, high performing teams.

Recalling the challenges that functional area allegiances can impose on team effectiveness, certain organizations have had success in overcoming this challenge through the use of a Matrix Structure. Matrix organizations are structured with specialists that are permanently assigned by project rather than by functional area. Cross-functional teams are totally different with people from different functional areas assigned on a temporary basis to teams.

MANAGING TECHNOLOGY DEPENDENT OPERATIONS

> *...maximize the prospects of the "end product" of their efforts...*

One impediment to the ability to rapidly configure teams in re deploying technical resources has been the dynamics associated with functional area allegiances. To counter this, the development of *cross functional* teams in which members from various functional areas are assembled together in a group focused on a common initiative have enjoyed considerable success in overcoming the functional allegiance syndrome. With team members bound by a common goal, the allegiance chord to their respective home functional areas are severed thus allowing team members to fully immerse and devote themselves towards the interests and goals of their respective cross functional team. The keys to success have centered on assembling members having unique skills and expertise within each functional area together in teams that are focused upon and solely dedicated to the objectives of the team. In structuring teams in this fashion, in addition to temporarily severing the functional area chord, organizations most proficient in the art and practice of Matrix Organizational Structures can go a long way towards building speed of configuring, deploying, and reconfiguring its scarce intellectual resources for competitive advantage.

Adjourning

> *...how best to handle the delicate balancing act...*

The *Adjourning* stage represents the organized disbanding of the team allowing its members to be freed up for redeployment on other initiatives. Recognizing the likely need for members to be able to work effectively and collaboratively together in future group assignments, disbanding should be orchestrated in a fashion that retains positive interpersonal relationships between members.

"By-the-book," once a team has completed its clearly assigned task and objectives, it should be adjourned to free people up for other critical

LEADERSHIP, MOTIVATION AND TEAMWORK
By Matthew Puz

initiatives. In some respects, success in TDO's is a numbers game – "How many differentiating initiatives can you drive towards fruition with the resources you have?" Yet, there may be special circumstances under which teams should be permitted to continue on for extended periods of time.

> *It is what reflects the planned team duration...*

What are these and what justifiable rationales exist for doing so? For example, teams given the charter of capturing and gathering technology-development efforts from disparate areas within an organization serve several useful purposes. Some of these include ensuring that technology-development efforts are not needlessly duplicated (re-inventing the wheel) and the transfer of promising technologies throughout the organization. In such cases, longer time horizons teams can be well justified, especially for larger organizations having a clear need to ensure that a high level of awareness about its technology development efforts exists.

Ensuring ongoing effective technology transfer, standing teams should be kept "fresh" by creating rotating membership amongst its members. Seasoned veteran members can serve as "mentors" by indoctrinating new, less experienced members into the desired cultural practices of effective technology development and transfer in the organization. On the other hand, bringing new members onto a team can provide fresh perspectives and a sense of renewal to the team – positives in both regards.

Team Performance: "In-Practice"

Few organizations can truly describe the working environments in which their active teams operate as being conducive to high performance? Many would say, "It's easier said than done and it won't (or doesn't) work in my company." So how can (tech) leaders get the rubber to meet the road en route to building, managing, and motivating effective teams? Consider the tools and techniques to follow.

MANAGING TECHNOLOGY
DEPENDENT OPERATIONS

On the Look Out for Sub-Par Team Performance: Far too often, protocols form within organizations in which sub-par team performance is not only tolerated and accepted, but becomes part of its culture. In these cases, teams often provide little in the way of measurable outputs leading towards the achievement of organizational goals. Especially in larger organizations, it is all too easy for teams and team membership to become simply a box to check with respect to organizational involvement, but having little "teeth" or responsibility related to actually delivering on results.

> ..."do more with less" in hyper-competitive times...

Needless to say, in the 21st Century organizations can ill-afford to have teams operate under these conditions so conducive to "underperformance." In these hyper-competitive times, it is crucial that organizations vulnerable to sub standard team performance proactively initiate steps to counter such situations by creating new protocols that heighten the performance standards of their teams. So where do we begin? Some approaches worth considering follow:

Setting a New Tone for Team Performance: In cases in which inferior team performance has become the norm, breaking the cycle of acceptance of such performance needs to occur. Doing so requires the setting of a new tone. In essence, a higher standard must be set as it pertains to expectations regarding team performance.

Technique 1, Current Teams Assessment: Check the feelings of team members. Are they up or down in spirit? Assuming they are on target with what you expect, do they feel that their team matters at organizational levels above? Do they realize that the performance of the team is viewed as being important by top management?

Technique 2, Escalating Top Management's Attention to Team Performance: As an antidote to a cultural operating environment accepting of sub-par team performance, increasing the attention paid to teams and their performance at the upper management level is critical. Paying greater attention to teams and their corresponding activities, progress, and performance will go a long way towards setting a new tone.

Technique 3: Selective Team Disbanding: In cases where the charter of teams are not viewed as being "mission critical," technology leaders should consider immediate disbanding them. In doing so, strong messages will be sent that a "turning of the page" is occurring and that teams and team performance are taking on a heightened level of priority. In doing so, two key themes will emerge – (1) "We're now keeping score on team performance" and (2) "Teams that may have existed as 'social circles' before will no longer exist."

In taking this step, managers must take care to guard against individual team members from disbanded teams being labeled as having "failed," - that is, unless a clear failure on the part of the team or specific individuals within the team has occurred.

Technique 4, Setting A New Tone: Consider having members of top management "sit in" on team meetings on an unannounced basis. Doing so will also send a strong message about the increased level of interest in and escalated expectations that the organization has of its teams. Additionally, the awareness that a member of top management may sit in on team meetings at any time serves to motivate individuals to ensure that they avoid coming up short in the "deliverables" expected of them by their respective teams. Having teams also periodically present the state of their progress in upper management staff meetings holds potential to further reinforce the desired messages about team performance importance and priority in the organization.

Technique 5: Purposeful Team Formation and Assignments:
1) Consider having functional area leaders also head up major projects. Doing so can create a "checks & balances" dynamic as functional heads become mutually dependent upon each other for resources

MANAGING TECHNOLOGY
DEPENDENT OPERATIONS

outside their home areas to support their project-specific responsibilities.

2) Consider formally announcing the establishment of new teams in a public forum to create heightened attention to, credibility of, and the importance of newly formed teams. In doing so peripheral areas of the organization, if called upon, will be much more likely to lend their support knowing that a high level of priority by top management has been placed on the project.

> *...the charter of each new team has been endorsed...*

3) Lastly, assigning persons considered to be "fast trackers" to high profile teams can also establish a positive protocol as it relates to career development opportunities. Doing so demonstrates that by contributing meaningfully towards high performance in a team setting as opposed to focusing on individual performance, one's career prospects can be better enhanced.

Consider the Skills Ballance: Although technical skills and knowledge are obviously important in TDO's, successful creation of high performance teams is also heavily dependent upon having a complementary mix of "soft" people skills to complement the "harder" technical skills.

> *...a "hard" versus "soft" skill balancing act ...*

In a perfect world, everybody within the organization would rank high in both technical ability and interpersonal skills. In the real world though, we all know that this simply isn't the case. Effectively addressing this requires that leaders ensure that a sufficient and appropriate level of members of having strong people skills are selected in the team formation stage to maximize the prospects of fully leveraging the harder "skills and knowledge" components present throughout the team.

LEADERSHIP, MOTIVATION AND TEAMWORK
By Matthew Puz

An assessment of the "personality mix" effect on team member compatibility should be done to promote positive interpersonal team dynamics. Potential

> *...use of an intervention session...*

pitfalls in this area relate largely to the "Forming" stage, where member selection is so critical, as during the "Storming" stage, where establishing proper interpersonal protocols are crucial.

Dealing with Team Discord and Poor Attitudes: Often unspoken hostilities hover over teams like a dark cloud and act as impediments to team efforts being truly collaborative. The destructive effects of poor attitudes on team performance may not be fully understood by all members. At times, overcoming team discord can seem to be almost insurmountable. Change intervention sessions designed to improve attitudes and promote open and honest communication can be tried. However, this is probably best done with the help of an organizational behavior specialist. The key is to get those harboring unspoken hostilities and resentment to openly articulate their "issues" as a crucial first step to overcoming such impediments to cooperative, positive team dynamics. Although temporarily it can create a high stress situation, in the end it can clear the air and create the opportunity for a "fresh start" to occur.

Addressing Non-Uniform Team Member Contribution: "Not everybody is pulling their fair share of the load." Let's face it. This is a common refrain. In certain cases, people want to contribute, but just can't seem to do so. In addition to team discord, poor attitudes, and the absence of essential "soft" skills, another impediment to team effectiveness relates to team members who can be loosely categorized as "prima donnas" and "gunslingers." These people tend to dominate their teams and/or portray a haughty demeanor – such individuals suck the life out of a team and drain enthusiasm from its members. A common result of their negative impact, in spite of their knowledge, is that less dominant (yet knowledgeable) team members "clam up." Again, at the Forming and Storming stages, leaders

MANAGING TECHNOLOGY
DEPENDENT OPERATIONS

> *The Prima Donna & Gunsligner Syndrome...*

need to be explicit that contribution from all members is expected. This imposes a requirement not only on introverts that struggle with making their voices heard to "speak up." but also on extroverts who tend to either consciously or unconsciously dominate group interaction. By guarding against this, ideas from all team members can be better fully nurtured, considered, captured and leveraged. In cases where non-uniform team contribution is occurring in your teams, consider the follow techniques.

Technique 1 - Team Facilitators:
The selective and purposeful placement of persons able to function as effective facilitators during team formation can go a long way towards ensuring that a sense of "balance" in group dynamics and contribution occurs. Again, people skilled as facilitators play a critical role in drawing out opinions, ideas and knowledge from all team members while also contributing to the creation of positive interpersonal protocols within the team.

Technique 2 - Post Meeting Electronic Debrief Sessions:

> *... avoid unrelated personal gripes...*

Conducting post meeting debriefing sessions to both summarize meeting minutes and action plans can provide a forum for extracting ideas from less vocal team members that may have been hesitant to offer the full breadth of their input during formal team meeting sessions.

Technique 3 - Team Member Removal:
As unpleasant as it may be, in situations where it is widely known that a team's poor performance is largely due to a person's lack of individual effort or negative attitudes, removal of a team member may be necessary. This highly visible action reinforces management's commitment to both uniform team member contribution and overall team effectiveness.

Technique 4 - Peer evaluations:
Another technique is to have team members do written peer evaluations (anonymously, if possible) assessing the level of performance and contribution of other members. This can flush out ineffective contribution and the existence of "bad apples." However, in doing so it is important to ensure that these evaluations don't become a forum for airing unjustified, personal gripes. Some keys include (1) Making sure that the team is large enough to allow for some degree of breadth of inputs and (2) Leaving room for you, as the manager in charge, to include your own observations. Guidelines to provide evaluators perspective in gauging their fellow teammates during peer performing peer assessments include:

- **Effort**
- **Attitude that impacts team dynamics**
- **Degree of tangible contribution related to skills and knowledge**

In certain cases, posting the results of peer evaluations or making them available for the organization at large to see upon request may be a good idea. The high performers may be reinforced having been publicly recognized. This is especially likely for the less socially active technical specialist of the team. The awareness amongst all team members that "slackers" or those possessing bad attitudes will most likely be exposed as having "short-changed" team members can act as a reprimand negatively reinforcing this type of behavior. In effect, a publicized peer evaluation ensures that their performance and contribution levels to their respective teams will not go unnoticed.

Technique 5 - Team and Individual Rewards:
Clearly, a strong argument can be made for constructing a reward system that creates a focus on group performance over individual performance. In doing so, people become more inclined to subordinate one's own interests to those of the group. A group reward system also creates a healthy dose of

MANAGING TECHNOLOGY
DEPENDENT OPERATIONS

positive peer pressure that goes a long way towards ensuring that any "deadwood" or persons creating unhealthy dynamics or not pulling their weight get "rehabilitated" or removed from the team. Again, intermittent peer evaluations provide a chance for identifying and addressing these types of concerns.

Techniques capable of addressing the origins and causes of "Cultures Accepting of Sub-standard Team Performance" and "Non-Uniform Team Member Contribution" are provided in the Appendix. As it pertains to the formation, management, and motivation of High Performance Teams in a general sense, Table 13 provides a summary of many of the key building blocks at the disposal of tech leaders pursuing enhanced team performance.

Table 13: High Performance Teams Checklist (7)

Key Success Factors	*Complete Yes / No*
1) Clearly identified team goals and performance expectations	
2) Appropriate skill composition profiles, Pre-screened team member fit for contribution potential in both "hard" and "soft" skill areas.	
3) Clear communication of individual roles and expectations regarding contributions to the team.	
4) Creating ownership & commitment by involving team members in shaping goals and allowing teams to determine best tactics by which to pursue goals. .	
5) Fostering a climate that enables the group to create momentum through the achievement of "early wins"	
6) Setting a tone of constructive interpersonal protocols for team interaction and behavior consistent with the culture and values desired in the organization.	
7) Providing adequate resources required and eliminating obstacles to team performance while granting sufficient autonomy for teams to be self-directing.	
8) Providing desirable rewards for both effective team performance and individual performance fostering a "team first" view as opposed to individually motivated.	

LEADERSHIP, MOTIVATION AND TEAMWORK
By Matthew Puz

The Leap to High Performance How does one take these concepts so crucial to team success to the next level – from commitment to devotion? The answer, at least in part, seems to lie somewhere at a point where leadership and motivation intersect in a team-based environment. We all know a pumped up team when we see it.

> *...greatness exists within people ... giving them opportunities ...to find and display it...*

It lies in presuming that potential for superior performance exists in all organizations' people and in giving them motivation and opportunity in team-based environments to find and display it. It also lies in spending lots of time with people and in relentlessly communicating and motivating them to scale greater heights. Through effective leadership, it is clearly possible to establish crucial emotional connections between group members linking them to their group's greater purpose.

Some of the magic in creating heightened team performance depends upon transforming the group from the "commitment" stage to the "devotion" level by touching the "soul" of the group. One technique for creating heightened levels of commitment and devotion to team performance is the *self-forming team*:

> *...from the "commitment" stage to a level of "devotion" through the self-forming team:*

Ideally, teams would be comprised of persons having uniformly high levels of commitment and self-imposed accountability to both their respective team members and to the achievement of its established goals. Towards this end, especially in larger organizations having deep talent pools, the concept of self-forming teams is worth considering. In such cases, once a general goal for the team is established and a small core of persons (champions) are in place, this team core is afforded the opportunity to

MANAGING TECHNOLOGY
DEPENDENT OPERATIONS

construct the balance of their team. Candidate selection methods include soliciting voluntarily membership or drafting up invitation or announcements to others who may feel themselves to possess the proper qualities (skills, knowledge, resources, attitude, team dynamics, commitment, etc.) to contribute meaningfully to the team.

> *...published peer evaluations and group rewards create a sense of higher stakes and increased ownership...*

One likely result of allowing teams to form themselves is that persons having well-established histories of effective performance in a team setting will be "in demand." Conversely, "deadwoods" are much more likely to be left on the sidelines. In this regard, the practice of Self Forming Teams, Team-based rewards and Peer Evaluations can reinforce each other.

The Team Leader and Key Roles Decision

For the case when a team is formed by edict, management should give serious consideration to who should be the team leader. By positioning strong technical contributors as the central figures in the teams being formed, it is possible to create a protocol that positively reinforces the organization's technology based goals.

> *...reward those most instrumental...*

"Authors" of promising technology concepts represent natural candidates to champion the respective technology development initiatives all the way to profitable commercialization due to the emotional attachment, ownership and commitment that they are very likely to have. They are much more prone to display vigilance in ensuring that projects stay "on track" and make meaningful progress. Having the authors of promising "technology seeds" hold central roles within the team's leadership structure sends a strong message: early pioneering efforts directed towards technology development or its new application will be rewarded with the opportunity to play crucial, high visibility roles on the teams whose formation comes about as a result of them.

Clearly, putting someone in the firing line in a key managerial position who lacks the necessary skills is not in the best interest of either the individual or the organization. However, training a key technical specialist with needed management skills should be a priority. If ill-equipped to formally lead such teams, "technology experts" could be "second in command" to a well seasoned team leader to bolster their managerial skills, while undergoing "on the job training" under such a mentor.

Large versus Small Company Distinctions and Opportunities

An organization's size represents a significant factor as to how much latitude exists in tailoring and implementing the tools to enhance the effectiveness of its teams. As the lines between organization's value (supply) chains become more and more blurred and increasingly co-dependent for technological support, a unique opportunity exists.

In the spirit of extended enterprises, individuals from smaller supplier companies serving as members of customer's development teams gain access to technology and the opportunity to gain new business. For such small companies, participation with larger companies that have well established leadership positions provides an opportunity to also be exposed to their "best practices" which can be brought back and potentially instituted into their own organizations. Large companies whose employees serve on technology development teams with smaller, successful organizations, are likely to learn

> *...exposure to successful entrepreneurial practices...*

MANAGING TECHNOLOGY
DEPENDENT OPERATIONS

some lessons in entrepreneurship. It can spark innovation. Successful entrepreneurial practices are often present in smaller organizations and should arguably be more sought after by the large ones.

Challenges of Team-Based Global Technology Management: The emergence of hyper-competitive conditions on a global scale has created the need for fresh perspectives about virtually every facet of their management practice. This has prompted several far-sighted companies to begin taking a global approach to their technology development and management. By establishing foreign-based technology centers or forming global strategic alliances, companies are positioning themselves to draw from the global pool of technology resources.

> *...draw from the global technology-development pools...*

Certain technology leaders have been trailblazing in the development of global technology teams. The use of Virtual Teams holds much promise. The power of the Internet age enables communication to occur either simultaneously by conducting meetings at pre-arranged times or sequentially through the use of threaded discussions in which people can offer their input electronically for all team members to view at their convenience.

Despite its benefits, the use of Virtual Teams presents challenges in the form of absence of face-to-face contact and the corresponding loss of communication through body language. Additionally, the benefits associated with team bonding that frequently occur when team members are in a common physical setting are lost. To overcome these limitations, intermittent face-to-face meetings and telephone conversations can be used to supplement virtual communication to impart a greater sense of "touch" to the interaction between global team members.

LEADERSHIP, MOTIVATION AND TEAMWORK
By Matthew Puz

On the positive side, recent research indicates that virtual meetings are particularly effective at soliciting participant feedback from international team members that may be uncomfortable with verbally expressing themselves in English. In a virtual setting the process is often able to more effectively extract valuable opinions and information that otherwise might not have been brought to light thus enhancing team member involvement and input (10).

> *...virtual meetings are effective at feedback from international teams that may be uncomfortable verbally expressing ...*

As the software in the form of "virtual tools" has gotten better, the ability to transfer information in an effective fashion has likewise improved markedly. With the luxury of enhanced information flow, companies are realizing the benefits of being able to cut down on travel for its team members thus simultaneously reducing costs as well as the "wear & tear" on their people.

Technology-leaders based in advanced countries are finding that a component of their future success will depend on their ability to effectively draw from the global talent pool. Doing so has become increasingly important as they continue to find their own talent and contribution levels being gauged against the "global brain-power" their competitors access.

As an example, General Motors has formed a cooperative R & D consortium, team effort involving their Powertrain and R & D Groups with a powerful network of Chinese universities and research institutes, as well as the Chinese government's science & technology sector. This has allowed them to tap into the knowledge bases of their Chinese scientist partners enabling them to

> *...acknowledged ...core technologies of tomorrow... may lie outside their own organization...*

MANAGING TECHNOLOGY
DEPENDENT OPERATIONS

bridge technology gaps. The benefits of taking such a bold step cut across multiple fronts. Tops on the list includes sending a strong message to the organization at large about the need to adopt a "global mindset" about technology while also capitalizing on the more tangible benefits associated with the adoption of global team approach.

Inspiration for Team-Building from Vietnam

On a personal level, a few years ago during a flight from Chicago to Dallas, I was afforded the opportunity to gain some unique insights about the essence of team effectiveness. As fate would have it, during this trip I happened to be seated next to a rather ordinary looking man in his mid-50's that I came to find out held some unique insights and perspectives. One might think that the man was a senior partner at a management consulting firm that was pushing his intellectual wares. Far from it – This was simply a regular guy named Frank dressed in summer attire flying off to visit his grandkids (presents in tow) proudly sporting a baseball cap identifying him as "Grandpa."

Despite my repeated body language intended to portray my strong interest in being left alone so as to allow me to prepare for an upcoming business meeting later that day, my companion's ceaseless persistence in wanting to engage me in conversation eventually wore me down and won me over. I reluctantly succumbed and, before I knew it found myself in the midst of one of the most enlightening conversations I would ever have on an airplane. I discovered that my travel companion was an injured Vietnam War veteran. Unlike multitudes of others that served there, his experience there was far different from the norm. The story goes something like this.

> *...efforts focused on creating an advantage*

In 1969, like thousands of others, Frank gets a draft notice in the mail. But upon reporting and checking him out, the Army soon discovers

144

LEADERSHIP, MOTIVATION AND TEAMWORK
By Matthew Puz

that Frank is talented with rifles. He's a great shot - a really, great shot. He soon gets the news that he's going to become part of what the Army calls their "Special Units" – Each comprised of "Specialists" having unique skills in any one of a variety of specific areas, one of these being marksmenship. So Frank goes into training to become a sniper as part of his Special Unit while other members prepare in the areas of mapping & tailgating specialists (for perimeter protection and movement recommendations), weapons/ explosives experts, communications and linguists experts (11). Of particular interest to me was the fact that in addition to their unique areas of expertise, each team member was cross-trained to provide them with a broad cross section of skills in each of key areas of expertise possessed by their respective team members.

I came to find out that the general mission of the Special Units was to expand the perimeter of the fighting forces at large. The object was to draw out the enemy, forcing their hands - in essence, to "throw the 1st punch." At the same time they were to extract as much information as possible about enemy activities, locations, and fighting plans. In this way, their efforts focused on creating an advantage for the American forces. The results of their efforts provided insights about the battlefield landscape that assisted the fighting forces at large in their ongoing advancement and attack strategies.

In the war many died. And further, the Vietnam War caused severe emotional damage on those who served and survived. Many exposed to the disturbing conditions and events of Vietnam are haunted to this day and are still suffering from the lingering emotional effects of the images they witnessed and events they were part of there.

And yet, incredibly, Frank's experience in 'Nam was still overwhelmingly "positive." This blew me away. How could this be? I came to find out that his experience was so positive that after completing his first tour of duty, Frank re-enlisted. He chose to go back into these conditions that some found so disturbing that they opted to take their own lives after returning home, even after they were no longer amidst these horrific conditions. Remember, this isn't a guy that was originally looking to serve his country. He was drafted.

There's more. Near the end of his second tour of duty, Frank gets hit in the head with shrapnel cutting it short. After Frank completed recuperation from his injury, shockingly, he approached the Army about returning for a _third_ tour of duty. The Army's response was "A guy with a shrapnel wound to the head wants to return for a third tour?" No way. Out of the question. Request denied."

> ... *"with a shrapnel wound to the head - wants to return..."*
> *...team spirit?*

Frank's reaction? "I decided that I wasn't going to take it lying down so I filed a formal protest to get them to allow me to return. I needed to get back there. I missed it." "What? You missed it? How in God's name can that be?" I asked. "How do you explain your insatiable desire to return to the horrific conditions in Vietnam that so many others had found so disturbing? That just doesn't add up."

What was it that created such a passionate fire within this guy that made him not only survive in these conditions that disturbed others so profoundly, but to actually thrive*? Given everything that I had heard and knew about Vietnam, I had to know what accounted for his experience there being so overwhelmingly positive and exhilarating.*

Recognizing my confusion and sensing my genuine curiosity, Frank nods his head, takes a breath, and begins: *"You have to understand, our team was different. Special. And we knew it. We were given a charter and were asked to fill it. Nobody told us how to get the job done. Nobody could. In the jungle, there's no time for second-guessing or bureaucracy. We were given a job to do and were trusted to get it done in whatever manner we determined it could best be done. Our group was elite and we knew that, too. So much so that we took orders from and reported directly to General*

LEADERSHIP, MOTIVATION AND TEAMWORK
By Matthew Puz

Westmoreland. We were making a real difference in the war effort and, without us and our contribution, the challenge of our fighting forces would have been even greater. We knew that in the absence of what we were doing, our troops would have been dramatically less equipped to fight the war. Where we treaded and made our impact felt, we stacked the deck for those that followed us to succeed. What we did really mattered. And further, we knew that because of the unique skills mix we had within our group that nobody else could have done the job that we were doing in the way we were doing it. There was something very powerful about that that's hard to describe. Can you begin to understand?"

At that point, it started coming into focus for me and I started to get it. His "I missed it (the war)?" comment began to hold some water for me. Then it dawned on me. Frank's story began to create some clarity for me as it relates to the business world. It dawned on me that if you can get a guy to literally fight for the opportunity to return to their team in the horrific conditions of Vietnam, it's got to be possible to create high performance teams in literally any organization under any set of circumstances.

As I recount my memorable conversation with Frank, I realized that there were some pretty powerful lessons for business people who are challenged to construct, motivate, and maximize the performance of effective teams within their organizations. The conclusions I reached about the effectiveness in the manner with which the Army's Special Units created and cultivated a team-based environment where team members thrived at both the individual and group levels boiled down to following:

> **Crystal clear goal clarity,**
> **Commitment to "the cause,"**
> **Complementary skill fits,**
> **Opportunities/expectations to bring out the best,**
> **Strong group bond & mutual dependence,**
> **Urgency to perform – High stakes**
> **Self-management / Autonomy,**
> **Expectation of high performance - "elite" status.**

MANAGING TECHNOLOGY
DEPENDENT OPERATIONS

This compelling example shows what team spirit can do built through effective leadership and motivation. Commit to stepping up to the challenge of tapping your organization's "software" for both its betterment and for your own professional development. It's a "win-win" proposition.

REFERENCES

1. Bernard M. Bass, ed., *Stogdill's Handbook of Leadership*, The Free Press, New York, 1981.

2. *Hersey-Blanchard Model, Management of Organizational Behavior, Prentice-Hall, Englewood Cliffs, NJ., 1988, P 171)*

3. *Time Magazine*: June 18, 2001, "Blind to Failure." Pgs 52-63

4. Mort Meyerson, Fast Company, *Handbook of the Business Revolution*, 1997

5. A.H. Maslow, "A Theory of Human Motivation," *Psychological Review*, vol. 50, 1943, pp.370-396.

6. V.H. Vroom, "*Work and Motivation.*" John Wiley & Sons, Inc., New York, 1964.

7. Katzenbach and Smith, "The Discipline of Teams." *Harvard Business Review*, (March/April, 1993 (pp. 111-120)

8. Heinen and Jacobson, "A Model of Task Group Development in Complex Organization and a Strategy Implementation." *Academy of Management Review*, Vol 1 (October 1976), pp.98-111

9. Irving L. Janis, *Groupthink*, 2nd ed. (Boston: Houghton Mifflin, 1982) and Irving L. Janis, "Groupthink," *Psychology Today* (Nov. 1971), pp 43-46

10. Alison Overholt *Virtual Teams – Fast Company*, March 2002

11. *USA Today,* "Next Stop: Special Ops." October 11, 2002

MANAGING TECHNOLOGY
DEPENDENT OPERATIONS

Suggested Additional Reading

- John R. Katzenbach and Douglas K. Smith, **The Wisdom of Teams: Creating the High-Performance Organization**, Harvard Business School Press, Boston, 1993.

- For additional reading on the subject of the impact of leaders time allocation, consult "What Effective Leaders Do", **Harvard Business Review**, John Kotter, 1985

- For additional reading, insights and data related to the subject of time allocation, consult Peter Druckers, Chapter 2 entitled "Know Thy Time" contained in his work "**The Effective Executive**", Harper & Row Publishers, 1966.

- Charles C. Manz and Henry P. Sims, Jr, **Business Without Bosses: How Self-Managing Teams Are Building High-Performance Companies**, John Wiley & Sons, New York, 1995.

- Dennis A. Roming, **Breakthrough Teamwork**, McGraw-, New York, 1998.

APPENDIX

Table A-1 Overcoming a Culture of Sub-Par Team Performance

Keys to Setting a New Tone	**Yes**	**No**	**Possible Action**
Current Teams Assessment			
Selective Disbanding of Non-Critical Teams			
Escalating Top Management Commitment to Teams ➔ Periodic Sit-Ins by Senior Executives ➔ Intermittent Team Presentations			
Purposeful Team Formation & Assignments			

Table A-2 Overcoming Non-Uniform Team Member Contribution

Keys to Setting a New Tone	**Yes**	**No**	**Possible Action**
Peer Evaluations			
Individual and Team Rewards			
Team Facilitators to Enhance "Soft Skills"			
Post Meeting Electronic Follow Up Sessions			
Selective Team Member Removal			

MANAGING TECHNOLOGY
DEPENDENT OPERATIONS

Table A-3 "People -Time Quotient" Tracking Table

Time	Monday	Tuesday	Wednesday	Thursday	Friday
Before 8:00 AM	I	I	O	I	L
8:00 AM	etc.	L	L	I	I
etc. etc.	etc.	O		L	I
12:00 PM	etc.		O	L	
etc. etc		L	L	I	L
4:30 PM	etc.	I	L	L	L
After 5:00 PM	etc.	L	I	O	I

Key Terms:

L = "Leading" Activities ➔ Collaborative Interaction

I = "Individual" Activities ➔ Working largely independently, absence of "Leading"

O = Other ➔ Clarify.

Note: Within each box, briefly noting the nature of the tasks associated with each time segment is recommended. Subsequent debriefing of one's day can provide valuable insights as to trends and tendencies that are occurring and impacting an individual's time allocation and corresponding leadership effectiveness.

Chapter 5, Lead-in:

MANAGING INNOVATION

Creativity in the arts is obvious. In research and engineering, technical creativity has been studied methodically and extensively. In operations, where efficiency, cost control, and steady-state is often the traditional goal, promoting and managing technical creativity has some unique challenges rarely discussed.

Successful management of technological creativity requires in-depth understanding of organizations' cultures. This chapter covers a wide range of perspectives on the expectations for technological creativity. Failure to achieve results of technically creative activities are more often related to differences in expectations than to failures in achieving technical goals. What a scientist in R&D and an operations or business manager sees as innovation often are not the same.

Discussed in the chapter is an innovation process map. It suggests opportunities for better understanding of the role of specific activities in the technological creation process. For example, idea creation ("ideation"), market research, and competitive intelligence to name a few.

MANAGING TECHNOLOGY
DEPENDENT OPERATIONS

CHAPTER 5

MANAGING INNOVATION

John M. Fildes, Ph.D.

Although a simple concept in an academic sense, management of innovation is a complex activity for technology managers. Innovation is multifaceted, but technology and business managers are poorly educated and unaware of the numerous perspectives they need to deal with. Technology and business managers are also not adequately trained in the tools that support innovation in a global environment. Innovation is an ad hoc activity in far too many organizations. Successful management of innovation requires in depth understanding of organizations' cultures, the outlooks and expectations of various functional areas, and a process that couples technological creativity to stakeholder needs and expectations at the front end and to engineering and manufacturing capabilities and limitations at the back end.

> *Innovation is a complex activity*

Innovation is a centrally important aspect of today's business environment. Although strategy, execution, and capitalization remain cornerstones of businesses, innovation is becoming an equally important cornerstone in many sectors. The entire Internet economy is based on innovation in business models, which are themselves becoming as important as products as defining features of many businesses. For example, e-Bay

MANAGING INNOVATION
By John M. Fildes, Ph.D.

has revolutionized the auction business model through innovations involving use of the Internet. Innovation in business processes is also a common strategy for gaining competitive advantage. For example, many companies such as GE and the start-up SupplierMarket.com are using internet-based innovations to revolutionize the procurement process and supplier chain management processes. Innovation in product design has also grown greatly in importance. A common strategy for success in product-oriented companies is to introduce superior products more quickly, allowing customization to better satisfy the demands of specific customers. Often these superior products are also less expensive. At its extreme, this strategy, which is called mass customization, allows each customer to customize an item to his or her tastes.[1]

> *...internet economy is based on innovation...*

Other chapters of this book have looked at the ways in which innovation is used in successful strategies to gain differentiation, competitive advantage, and the associated higher margins. This chapter looks at the management of innovation. It offers business leaders and managers important knowledge, techniques, and tools to improve their management of innovation. The techniques and tools are demonstrated with real examples of their use. Although the discussion and examples are centered on managing product innovation, the techniques and tools are also readily usable for managing business model innovation. These tools and techniques are equally pertinent to large and small companies and to OEM (original equipment manufacturers)corporate managers, design service providers, and consultants.

Understanding Innovation

Innovation can occur in all activities and functions, including business models, strategies, processes, and product design as was discussed in the

---

[1] James H. Gilmore, Joseph B. Pine, "The Four Faces of Mass Customization," *Harvard Business Review*, 75(1), Jan/Feb., 1997.

MANAGING TECHNOLOGY DEPENDENT OPERATIONS

> *Innovation involves "introducing into use"*

introduction.[2] Innovation is often looked upon as a purely creative activity that cannot be precisely understood and described let alone managed. It is no wonder that there is a diversity of outlooks of what constitutes innovation in many organizations. Some view innovation as an act, which is incorrect. In the context of business models, strategies, and processes, innovation is often through of as the creation of ideas, which is also incorrect. In the context of product design, innovation is often confused with invention. Innovation involves "introducing into use" in most contemporary definitions. The generation of ideas and inventions are not innovation because by themselves they may or may not lead to a business result. Thus, innovation needs to be treated as a process and not as an act if innovation is to be managed properly. This is the central theme of this chapter, which introduces the concepts of the innovation process.

> *...link accountability with the creation of ideas.*

Recognizing innovation as a process is critical to addressing many of the complaints about the lack of creativity and innovation in organizations today.[3] Most businesses have an adequate supply of creative people, but they lack the processes to harness this creativity. Companies do many things that unknowingly inhibit innovation because too much focus is placed solely on the act of creating ideas or concepts and not enough emphasis is placed on thinking them through and instituting actions that bring them to reality. It is wrong to divorce the creation of ideas from the accountability for working them through to being discarded or acted upon. Fewer ideas which are followed through to their logical completion are far more valuable that numerous ideas that receive virtually no follow-up. Following through on ideas is natural in organizations that link accountability with the creation of ideas.

---

[2] Joan Magretta, "Why Business Models Matter," *Harvard Business Review*, May, 2002.
[3] Theodore Levitt, "Creativity Is Not Enough," *HBR*, August, 2002, pp. 137 to 144 (from HBR 1963)

MANAGING INNOVATION
By John M. Fildes, Ph.D.

> ...*leaders can effect innovation if they take ownership of the idea*...

The nature of the problem with not linking accountability with the creation of ideas is well exemplified by the common misuse of ideation and brainstorming. Ideation and brainstorming are often treated as autonomous acts rather than as part of a broader, well-defined process with a specific objectives and an endpoint. There is no definition of ownership, no clear expectations, and no definition of accountability for follow through until the ideas are discarded or implemented. This situation has numerous consequences that have a substantial impact. Ideas that do not have ownership and follow-up pose a severe problem to managers whose work lives are already overburdened. These managers view the ideas as a problem that adds to their workload. The ideas do not go anywhere, which leads employees to be skeptical of management's desire and ability to innovate. This culture of skepticism becomes ingrained and inhibits innovation.

Smaller companies are often viewed as more innovative than larger ones. This situation does not exist because smaller companies have more innovative employees. In fact, larger companies probably have an equal proportion of innovative employees so they have more in absolute numbers. Larger companies also have more resources to support innovation. They should be more successful at innovation. Rather, smaller companies appear to be more innovative because they lack the conformity mechanisms found in larger companies that inhibit innovation. In small companies, the founder or a group of closely-knit leaders maintain the culture through direct contact with all of the staff. These leaders can effect innovation if they take ownership of the idea. This becomes impractical as organizations grow beyond the size in which direction from a single individual or closely-knit group of leaders to the staff is practical. Part of the growth process for organizations is the necessary definition and institutionalization of beliefs, principles, and approaches, but this is in conflict with the requirements to have an innovative organization. Innovation requires change. Failure to treat innovation as a process that involves ownership by people with sufficient authority dooms ideation and brainstorming to be dead-end activities.

**MANAGING TECHNOLOGY
DEPENDENT OPERATIONS**

Understanding Innovation as a Process

The specifics of innovation processes will necessarily differ from situation to situation and company to company. Nonetheless, a framework that contains all of the functions that are common to all innovation processes can be defined as follows:

- Deep and specific understanding of customers' needs and business strategies.
- Clearly stated objectives with a specific outcome.
- Definition of each participant along with their accountability and roles.
- A logical set of activities with a timeline.
- Procedures and associated tools.
- Training.
- Ownership and leadership.
- Reporting.

For example in product development, understanding innovation as a process starts with the recognition that the innovation process involves customers, and possibly their customers, as well as the innovator's employees and vendors. It is important to understand and define the role of each participant in the process. It is also important to recognize that a consequence of this situation is that managing innovation will cross organizational and geographical boundaries. An effective innovation process must take integrating different cultures and geographical locations into account. This means that virtual organization management techniques and internet-based communications techniques and technologies will be important components of managing the innovation process shown in Figure 1. The innovation process for product design also involves integrated product design, engineering, process design, and rapid prototyping. This situation challenges technical managers in ways that they are not used to and for which they are not prepared.

MANAGING INNOVATION
By John M. Fildes, Ph.D.

Strong, Ongoing Customer
Interaction and Use of IT and Agility

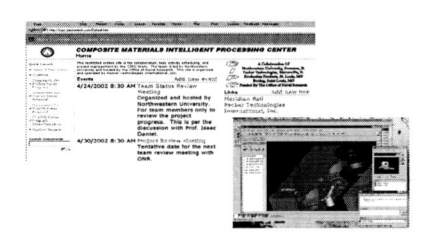

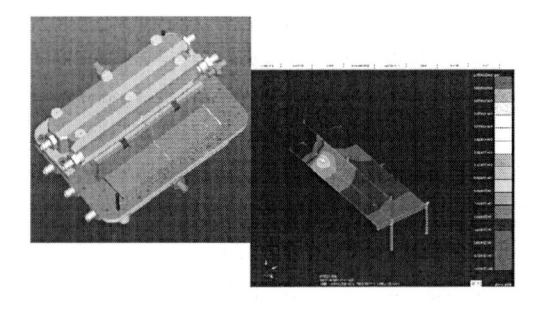

Integrated Product Design,
Engineering and Process Design

Prototyping, Testing

Figure 1: Components of Innovation in Product Design

The Role of Customers

Innovations in product designs are driven by cost-reduction and customization. This is easy to understand. Computer technologies in the form of design tools and marketing/sales vehicles are leveling the playing field. These technologies allow modestly capitalized start-ups to challenge and sometimes displace large, well capitalized, long-standing players in a product category. Computer design tools and marketing/sales vehicles commoditize product sectors, making value-added status hard, if not impossible, to achieve based on performance superiority or customer access. This forces companies to continuously cut costs and to differentiate themselves more and more by transforming a better understanding of customers' needs and desires into a steady stream of customized products and by using their knowledge to provide superior customer service.[4]

> *...innovation is being driven rather than acting as the driver.*

This environment has forced a change in the role of innovation in companies. Innovation used to drive product development, whereas in today's environment product development is driving innovation. Although it has always been important to manage innovation, it becomes far more important to do so when innovation is being driven rather than acting as the driver. Today's business leaders and managers are adjusting to this paradigm shift, and their success may make or break their companies.[5]

Customers need to be an integral part of developing innovative products. The traditional iterative process of producing prototypes, market testing, and revising the design is too lengthy. Customers need to be an ongoing, integral part of the design team, participating in decisions about functionality, form, performance, and cost. Innovation in product design also requires a much better understanding of customers' needs, opportunities, and limitations. This is especially challenging because customers frequently have not thought through their product strategy to the extent that ensures successful acceptance by the marketplace.[6]

> *Customers need to be an ongoing, integral part of the design team...*

It is very infrequent that the design of innovative products fails for technical reasons. Designs usually perform to the degree that was expected. What is far more common is for the customers' expectations to

---

[4] Nicola D'Amico, "A Measure of Success," *Business*, November, 2001.
[5] W. Chan Kim and Renee Mauborgne, "Knowing a Winning Business Idea When You See One," *Harvard Business Review*, Sept.-Oct., 2000.
[6] Anthony W. Ulwick, "Turn Customer Input Into Innovation," *Harvard Business Review*, January, 2002.

MANAGING INNOVATION
By John M. Fildes, Ph.D.

not be met. This can be because the customer's expectations were not appreciated by the design team or because the customer did not completely or properly think through what they needed. Design failures for these reasons are all too common and often occur because the role of the customer is not properly defined. Although, the customer needs to define what functions, performance, and cost are needed, the nature of the interaction between the designer and the customer is often one in which there is far more of the customer's focus on how the requirements will be met than on defining the requirements correctly.

Defining the requirements should be the role of the customer. Deciding how to best meet the requirements should be the role of the designers and engineers. Commingling these roles usually leads to very poor definition of functions, performance, and cost. It also sometimes leads to exclusion of valuable design features such as certain materials and processes. The result is that the prototype meets the performance that was expected, but it does not satisfy the requirements for the product to be successful in the marketplace. Revisions and additional prototypes are needed. There is not only a substantial impact on cost and the development time, but the ability to customize products for specific market segments is severely hampered.

The Role of Internal Staff

Dealing with innovation often exceeds the experience and training of the staff, yet many organizations provide no training and do not make the demands and roles of the

> *...provide on-going support in risk taking...*

innovation process a continuous topic of discussion. Innovation's increasing importance is forcing more interaction between scientists and engineers and non-scientists and non-engineers. These groups have very different training, thought processes, and terminology. Organizations need to provide on-going support in risk taking, accountability, and team building for scientists

MANAGING TECHNOLOGY DEPENDENT OPERATIONS

and engineers.[7] How many companies provide basic business and project management training to entry level scientists and engineers? The answer is that it is not the norm, but it should be. Non-scientist and non-engineers also need training in scientific and engineering methodology. Although science and engineering are challenging to learn to a degree sufficient to practice them, understanding their issues and approaches can be readily learned by non-scientist and non-engineers. How many non-scientists and non-engineering technology managers in your organization have undertaken any training to understand science and engineering methodology and management? How many companies offer this type of training?

Scientists and engineers are often different from non-scientists and non-engineers. They have endured a very challenging technical education and are most often very intelligent, but they may not be very effective in getting things done, which is a critical requirement of innovation. Scientists and engineers also tend to want to understand why specific demands are being placed on them. Questioning, challenging, and recreating/validating is the core of their education. It's natural for scientists and engineers to question what they are told. It's natural for them to rework the issue. They have been trained that this leads to advancement that it is good. This is part of the internalization process for scientists and engineers, but it infuriates non-scientists and non-engineers. They may view this behavior as uncooperative and non-trusting.

Scientists and engineers also differ from each other in important respects.[8] Scientists are usually more creative and less detail-oriented. Science education is strong on developing analytical skills and theoretical understanding. Engineering education is strong on developing experience in applying practical knowledge.

---

[7] John K. Borchardt, "Risk Management in Product Development," *Today's Chemist at Work*, April, 1999.
[8] Simon Ramo, The *Management of Innovative Technological Corporations*, Chapter 7, Wiley and Sons, New York, 1980.

MANAGING INNOVATION
By John M. Fildes, Ph.D.

Scientist and engineers are not trained in team building. They think of innovation as an individual activity and as an event rather than a process conducted by a team. In fact, their training is geared toward individual accomplishment, especially for scientists and engineers with advanced degrees.

> *...engineers are not trained in team building...*

Scientists and engineers are trained as if their careers will be in an academic-type environment, one in which technical considerations are paramount. They have little, if any, training in understanding consumer expectations, corporate organizational structure, organizational management, project management, risk-taking, entrepreneurship, and understanding basic financial reports.

Scientists and engineers also are trained to be certain of facts before making a decision. Although this may be highly desirable, it is not practical in making business decisions. This orientation inhibits risk-taking, which is essential to innovation. This area will always pose a significant challenge. The reality is that a detail-oriented, anti-risk-taking orientation is necessary at many times, but may be hard for non-scientists and non-engineers to understand. To better appreciate this, consider what qualities are important in the scientist who developed the metal alloy and the engineer who designed the structure of the plane or bridge you are sitting on.

Scientists and engineers also generally have no training in project management. They have had to develop their organizing abilities on their own. The results may be limited and unusual. They may find it difficult to breakdown a complex activity into

> *...lack of basic project management skills can be extremely frustrating...*

a simple set of tasks with an appreciation of their sequencing and interaction. This lack of basic project management skills can be extremely frustrating to non-scientists and non-engineers. Don't scientist and engineers get it? The answer may well be no. They have not had the training and experience that non-scientists and non-engineers have had.

**MANAGING TECHNOLOGY
DEPENDENT OPERATIONS**

The interface between technology and business needs far more attention in organizations. People who can bridge this divide are rare and should be sought and retained with great appreciation. Every innovation team involving technology needs to have bridging this gap as a defined role for an individual, and this needs to be viewed as a critical role on the team. The process of innovation is about people working together in a seamless way, practicing common approaches, analytical processes, and decision making skills.

The Role of Vendors

Vendors play a central role in the product development process and must be viewed as important members of the team. There are several drivers that have increased the importance of vendors. One is that as companies define their core competencies they tend to outsource functions that fall outside of their core. Another factor is the increasing variety and sophistication of prototyping and manufacturing technologies. It is increasingly impractical for companies to invest in and support all of the functions that create cutting-edge products.

> *...define core competencies ... outsource functions that fall outside...*

Having to cross organizational boundaries to form product design teams poses additional demands that scientists and engineers have not been trained to address. Organizations differ in their cultures, their approaches to risk and decision making, and their procedures to name a few areas that are important parts of innovation. There are also the logistics issues. The people who have to work together are not co-located and usually do not share the same resource management systems.

This situation is best addressed through virtual enterprise techniques. The virtual enterprise turns the design team into a fully functioning enterprise that comes together, performs the design task, and disbands for redeployment to the next project.

MANAGING INNOVATION
By John M. Fildes, Ph.D.

A Process Map for Managing Innovation

Since innovation is a process, managing it is similar to managing other processes. In addition to defining objectives, roles, and accountability, a flowchart of the activities should be developed and followed. The flowchart captures the best practices in the activities from throughout the organization, allowing all to benefit from them. It also provides a framework for training new employees and for gaining consistency.

> *...capture all of the essential activities...*

Any process for managing innovation needs to capture all of the essential activities to turn an idea or concept into a result. Whether managing business model innovation, strategy innovation, process innovation, or product design innovation, these activities in general include:

- Establishment of the true functional and performance requirements,
- Development of a plan of action, including the assignment of tasks and responsibilities,
- Execution of the set of actions,
- Monitoring and periodic reporting and reviews,
- Testing and validation of the results.
- On-going collaboration.

Figure 1 above showed the components of the innovation process for product design. The components are collaboration tools, integrated product design and engineering, and rapid prototyping and testing. Coupled with clearly defined objectives, a timeline, and the definition of the roles and accountability of people, the entire innovation process is well described in a manner that can be taught, monitored, and managed.

MANAGING TECHNOLOGY DEPENDENT OPERATIONS

The establishment of such a process will be described for product development as an example. Although specific to product development, this example shows the steps, approaches, techniques, and some of the tools that are pertinent to all innovation processes.

The innovation process map for product design is shown in Figure 2. This process captures and defines all of the activities that are critical to innovation in product design. It starts with establishment of requirements; progresses to integrated design and engineering activities; and ends with prototyping, testing, and validation. It captures the flow and interrelationship of the activities that form the innovation in the product design process. This is an actual innovation process used by a startup product design venture. This innovation process consistently removed 5% to 40% from the cost of near-net-shape plastic and cast metal products and shortened the design time by 20% to 50%. For example, metal castings for large complex components were routinely designed in 50% of the time traditionally achieved and 25% to 40% cost reduction was routinely achieved.

Requirements Finalization – This beginning step of the process might also be called the external or customer kick-off meeting. The process as shown assumes that there has been substantial prior consideration of the project. Prior discussions would normally include market assessments of benefits and price, one or more product concept meetings, ideation and brainstorming leading to preliminary design concepts, preliminary consideration of materials and manufacturing processes, preliminary consideration of prototyping methods and validation testing, and a proposal with a work breakdown structure, schedule, and cost estimate. In many ways, the requirements finalization meeting needs to be a reconsideration of all of this prior activity.

A design guide is a tool for coordinating the product development...

A design guide is a tool that is handy for coordinating the product development process. The design guide initially contains the customer's requirements, goals, and expectations which should be clearly defined as part of the requirements finalization process. All members of the team will have a copy of the guide which will be updated to include all design decisions and revisions, the conceptual or

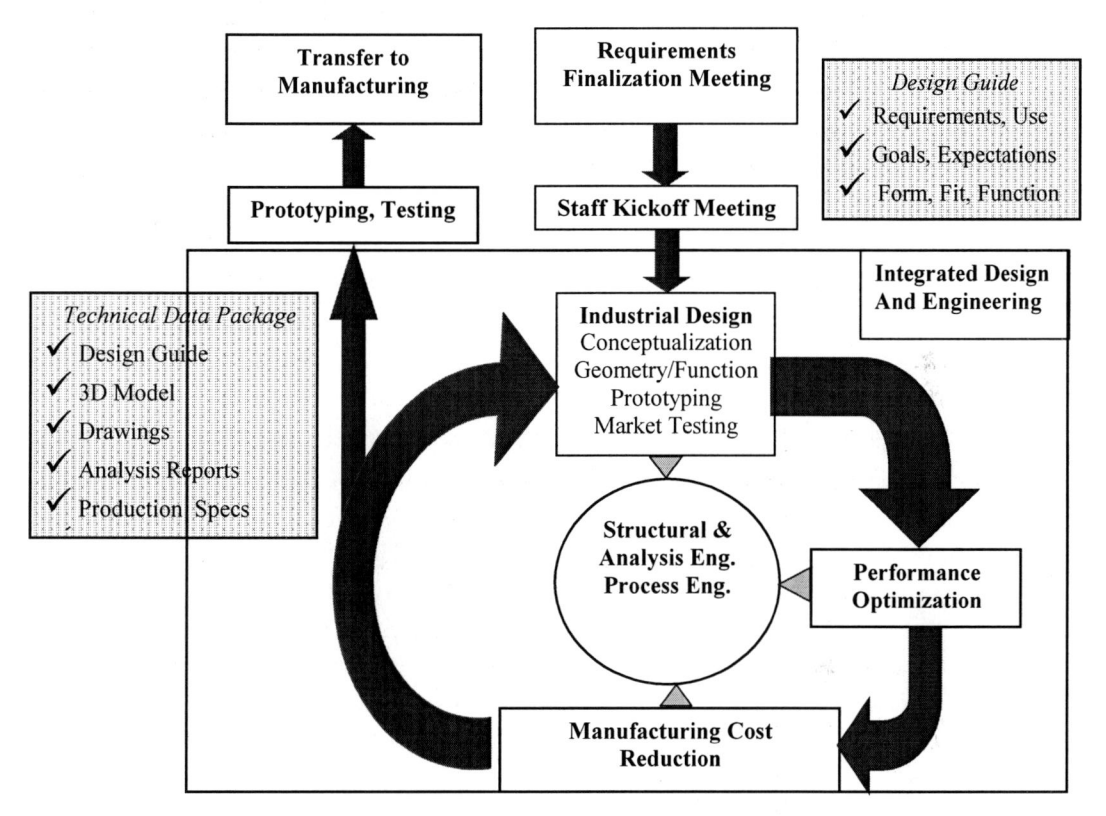

Figure 2: An Innovation Process for Product Design

functional design, the descriptions of the components or modules that make up the product, and the interface requirements for the modules to be brought together. These items will be further described as the remainder of the design process is described. The logistics of maintaining the guide, such as version control, will be further discussed in the section on internet-based collaboration.

MANAGING TECHNOLOGY DEPENDENT OPERATIONS

Staff kick-off Meeting – Managing the timeliness and quality of project work is essential for survival in today's competitive product development environment. The outcome of the design effort determines if customers come back and establishes a reputation in the marketplace. The goal has to be 100% customer satisfaction in today's environment. Managing the timeliness and quality of design project work requires recognition of the importance of these issues by engineers and their dedication to the daily use of the procedures and tool that are provided. This is a make or break issue for companies and the outcome is totally within their control.

The staff kick-off meeting is important so that every member of the team understands the customer's expectations, their role, the role of the other members, and the work plan. The design team leader should prepare a milestone chart and a work plan prior to the staff kick-off meeting.

The milestone chart is an important tool. Milestone charts contain a very succinct itemization of milestones (completion of critical path activities, deliverables, etc.) by month. Project leaders enter the planned milestones at the outset of the project. These are not changed without agreement by senior management. The actual accomplishments are entered at the end of each month. This information is used by senior management to monitor overall progress on satisfying the customer's expectations and the deliverables on each project. Project milestone charts are a much better tool than work plans for senior managers to monitor a portfolio of design projects. It is difficult to assess the information in work plans without detailed knowledge of the project. Planning the milestone chart also helps the project leaders and the project staff to appreciate the customer's expectations and the timing of critical activities.

> *...milestone chart is an important tool...*

Every project at its outset needs to have a work plan with a sufficient level of detail to manage it properly. Project work plans are for the project leader and staff to understand and coordinate their work assignment. Project work plans are sometimes overlooked by design team leaders in the mistaken belief that the Gantt chart in the proposal is an adequate schedule. This is seldom the case. Proposal schedules are intended for the customer to understand the sequence and timing of the logical steps to design the product. The level of detail that is appropriate for this purpose is usually not sufficient for leading the project. Also, the proposal schedule does not contain staff assignments and other information that is needed to manage the project.

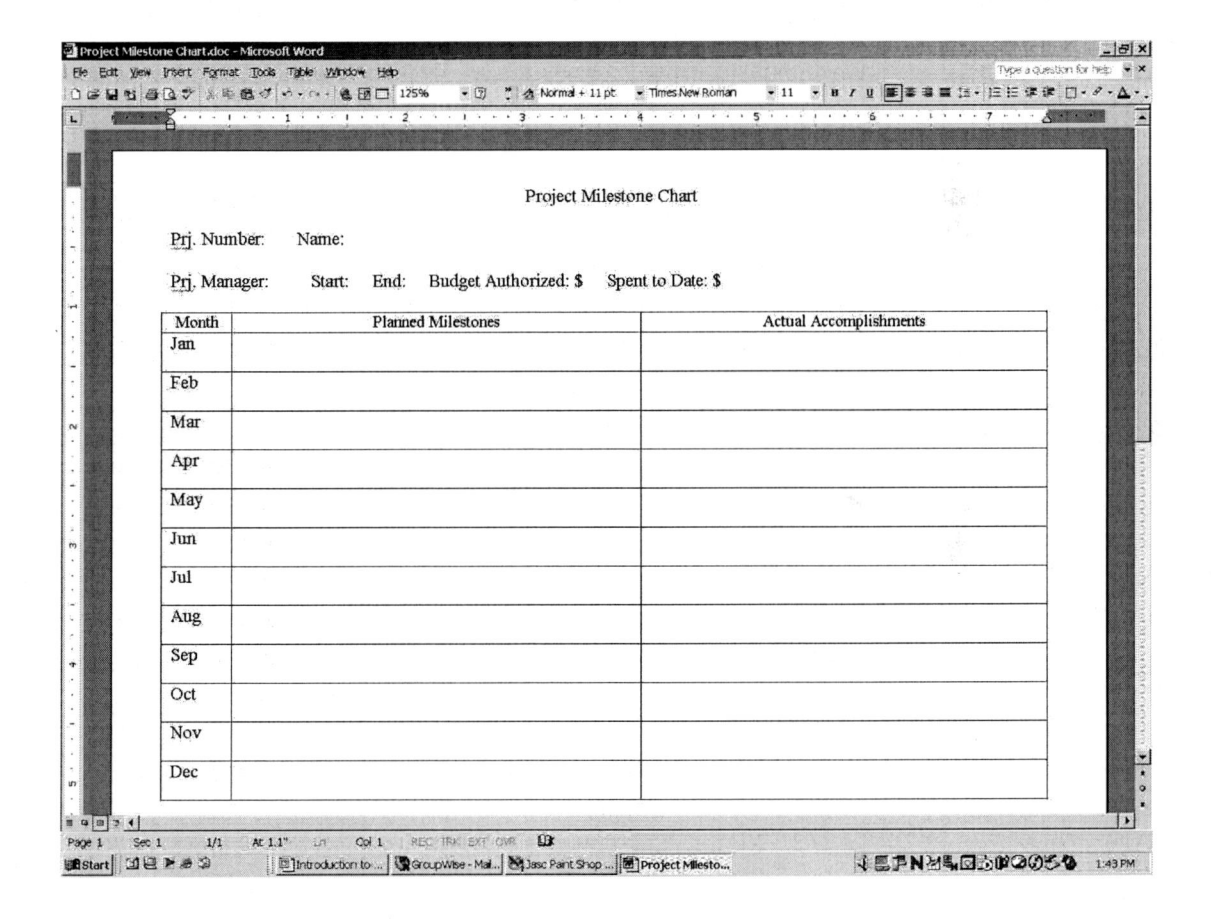

Figure 3: Design Project Milestone Chart

MANAGING TECHNOLOGY
DEPENDENT OPERATIONS

<u>Industrial Design and Engineering (Design, Engineering, Analysis, Optimization, Manufacturing Cost Reduction)</u> – These are a set of partially-concurrent and partially-iterative activities that are the guts of the design project. The design process has been revolutionized by the use of analysis and simulation tools, Figure 4. First, a three dimensional (3-D) solid model is developed. Then, a mesh is defined in preparation for finite element analysis. Finite element analysis allows important properties of the part to be calculated and assessed, for example stress levels and thermal properties. Other aspects of performance may also be calculated, Figure 5.

A recent innovation in the design process is the inclusion of process design. Often called design for manufacturing, integrated process and product design ensures that the design can be manufactured. It also allows the cost of manufacture to be minimized and the design to fully benefit from the capabilities of the process.

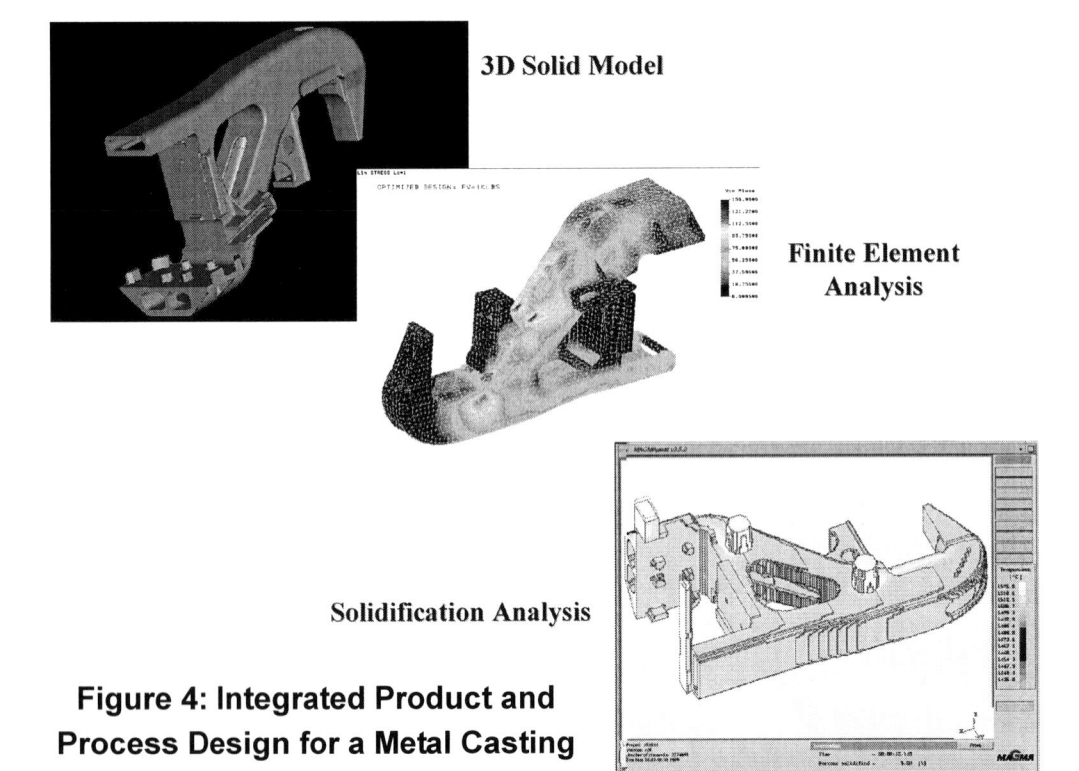

3D Solid Model

Finite Element Analysis

Solidification Analysis

Figure 4: Integrated Product and Process Design for a Metal Casting

MANAGING INNOVATION
By John M. Fildes, Ph.D.

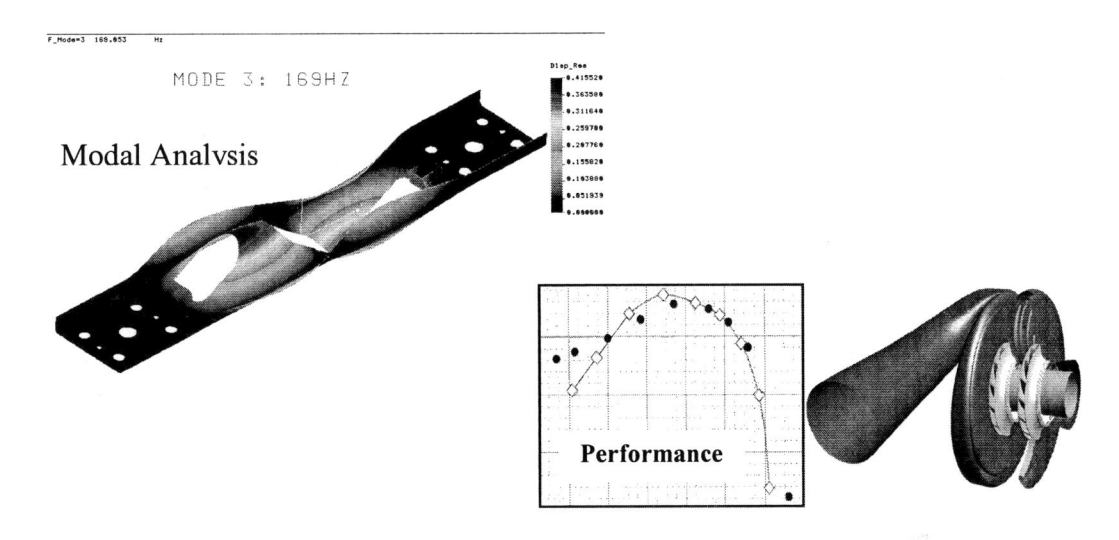

Figure 5: Performance Analysis

Prototyping and Testing – The purpose of analysis and simulation is to minimize the need for prototyping. Nonetheless, the design needs to be validated through prototyping and testing. Ideally, the use of analysis and simulation will result in the first design performing properly without need for adjustment or modification.

> *...design needs to be validated through prototyping...*

Selection of a prototyping technique is very important. Sometimes the actual manufacturing process will need to be used, as is often the case with advanced processes such as metal casting and composites fabrication. In other cases, injection molding for example, a process that is different than the actual manufacturing process can be used. This is advantageous especially when one of the rapid prototyping processes can be used. Rapid prototyping allows a part to be made directly from a solid model. An example of a rapid prototyping process is the laminated object manufacturing (LOM) process described in one of the case studies presented below.

MANAGING TECHNOLOGY
DEPENDENT OPERATIONS

Transfer to Manufacturing – The final step in the design process is the transfer of the design to manufacturing. This involves make-buy decisions and the qualification and selection of vendors. Part of this activity occurs early in the design process, during the design for manufacturing phase. Part of the activity occurs at the end of the design process.

> *The final step ... the transfer to manufacturing.*

Sometimes the design team will make this transfer, but often a customer has hired an independent design firm and the customer will make the transfer. There are important advantages for the transfer to be made by the design team.

Coordinating and Managing the Design Process — The design process shown in Figure 2 appears to be a relatively straightforward and linear process. This is not the case. There is much activity that underlies each of the steps in the process. This is shown in Figure 6. It may be valuable to flow chart the activities that compose each of the steps of the innovation process.

On-going collaboration is an essential part of the design process. This collaboration spans organizations and different geographical locations. There are many logistic and project-coordination issues. One example is ensuring that every team member has the current version of the design manual, which might be revised on a daily or more frequent basis by various members of the team. Internet-based collaboration and coordination address many of these issues. Nonetheless, internet-based collaboration is still unused or new to many organizations. This section covers both the techniques and tools. Although the technology and tools will change, the procedures will endure.

MANAGING INNOVATION
By John M. Fildes, Ph.D.

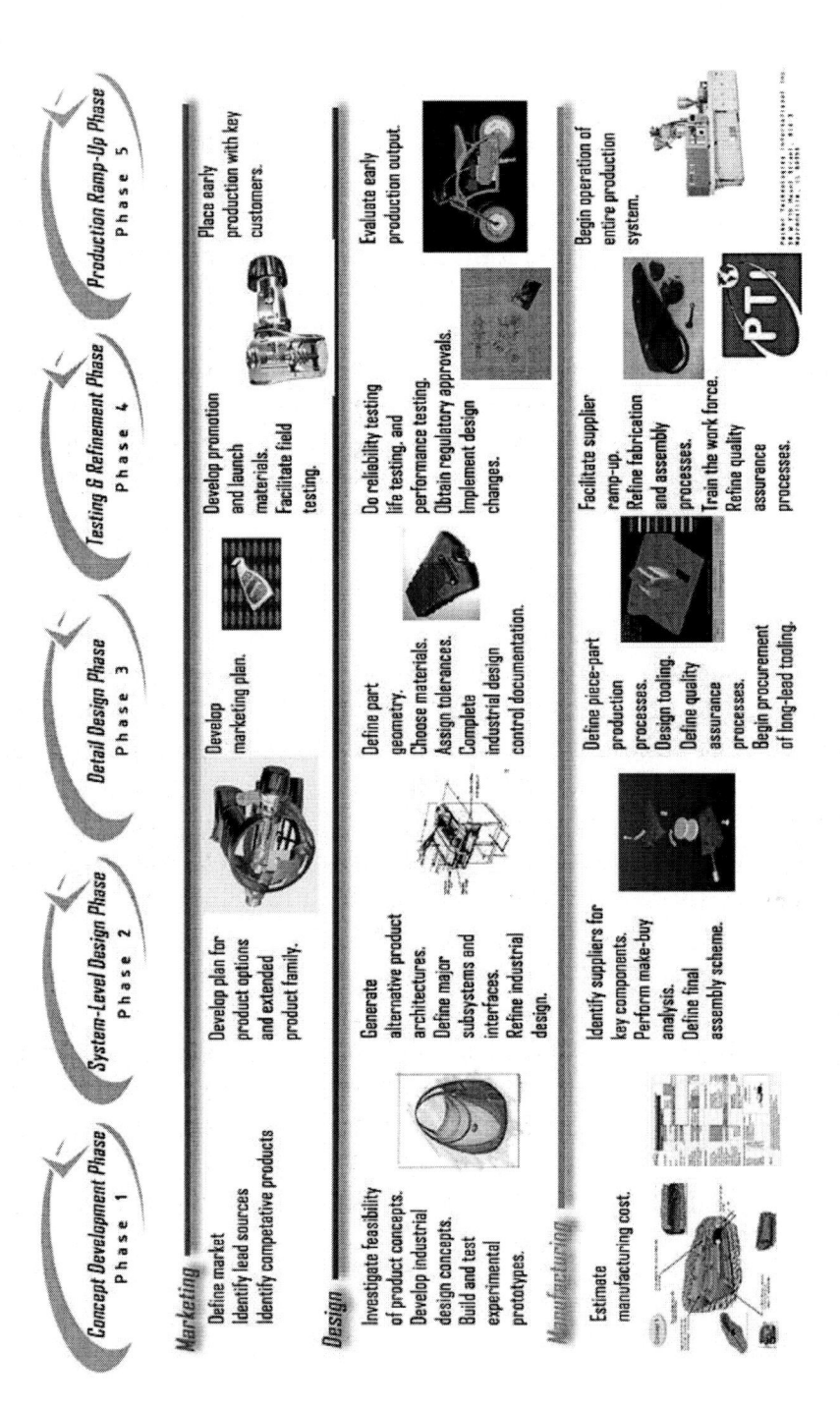

Figure 6: Functional Interaction in the Design and Engineering Phases

MANAGING TECHNOLOGY
DEPENDENT OPERATIONS

> *Internet-base collaboration offers a host of functions...*

Internet-base collaboration offers a host of functions that are useful for managing organizationally and geographically dispersed teams. These functions are:

- Password protected web sites
- Contact information
- Events list
- Task lists
- Document libraries with document version control
- On-line collaboration with whiteboards and program sharing

Internet-based collaboration software allows geographically-dispersed teams to share documents with assurance that they are using the current version. It also provides a way to schedule and monitor the status of tasks, schedule meeting, and have access to current contact information. The team can also hold virtual meetings, sharing presentations and engineering documents.

Internet-based collaboration software also contains tools that allow design teams to employ new techniques in the design process. For example, threaded discussion groups are common on the Internet, being used for interest groups, such as photographers, to share information. Threaded discussion groups can be used in an innovative way in the design process to reach consensus on functions, features, and design options. Another feature, document discussion, can be used in an innovative manner to edit reports.

Several different software packages that are useful in the design process exist. These packages offer similar core features as described above. Some offer additional specialized functions, such as supplier-chain management and on-line quote solicitation. Most packages are offered as a service with a monthly subscription fee. At least one is offered to be installed on a user's server and operated by the user.

Setup of these software packages is generally straightforward but may require experience with setting up and administering a server. Some of the packages are accessed through commonly-used Internet browsers. These packages offer simplicity of setup and use. Other packages require that a client program be installed on the user's computer. These packages can require special setup of firewalls that may not be acceptable to some organizations.

Case Study

Figure 7 shows the homepage of a password protected collaboration website. This site is for a Composites Intelligent Processing Center that is funded by the Office of Naval Research, managed by Northwestern University (Evanston, IL), and co-managed by Packer Technologies International (Naperville, Illinois), which also is the design member of the team. The team also includes Boeing (St. Louis) which is the customer, Production Products (St. Louis), which is a composites fabricator, and a Navy laboratory, which is the customer's customer.

The Center's mission and operation are an innovative approach to addressing the limiting problems that are actually faced by RTM manufacturers such as Boeing and Production Products. The manufacturers define the problems. Solutions are devised by Northwestern, Packer, and several Navy labs in collaboration with the manufacturers. Evaluations and demonstrations of the solutions are performed in actual manufacturing environments by the manufacturers.

This innovative approach overcomes a number of problems that are encountered in developing and deploying new technology. The approach makes sure that the people who experience the problems every day are the ones that define them. Technical people who are not on the manufacturing line each day tend to define the problems more on perceived technology gaps than on actual experience. Also, development of technical solutions can produce results that are not easy to use for manufacturers. Technology

MANAGING TECHNOLOGY
DEPENDENT OPERATIONS

will not be used if it is not easy to use, or if it requires discarding the recently-started corporate initiative, or discarding the large capital investment that has not been recovered. Demonstrations of technical solutions that are done outside of the manufacturing environment do not really validate the results because they generally use more capable people, equipment, and facilities than exist in the manufacturing environment.

Although the approach to the Composites Intelligent Processing Center addresses many problems that impede the development and use of new technology, the approach poses new problems related to organizing and coordinating a team that came from different corporate cultures and geographical locations. Utilization of virtual enterprise formation and operation techniques and internet-based collaboration were the answer to these barriers.

The website shown in Figure 7 was created with Microsoft's Share Point Team Services by Packer Technologies International (PTI). Hundreds of websites can run on a single server. PTI operates dedicated websites for a number of its customers. Share Point Team Services also allows sub-webs, so each team member in a multi-member program, such as the Composites Intelligent Processing Center, could have its own password protected sub-web. This is a very flexible arrangement for structuring a collaboration website.

The website is accessed through commonly-used Internet browsers so there is little training needed for users. The homepage offers easy access to all of the site's areas and functions. An event list is used as a reminder for meeting and reports. An announcements list allows the members to disseminate information of interest to the team. A contact list makes it easy to find phone numbers and e-mail addresses for all of the team members. The links list allows easy access to each member's own website.

MANAGING INNOVATION
By John M. Fildes, Ph.D.

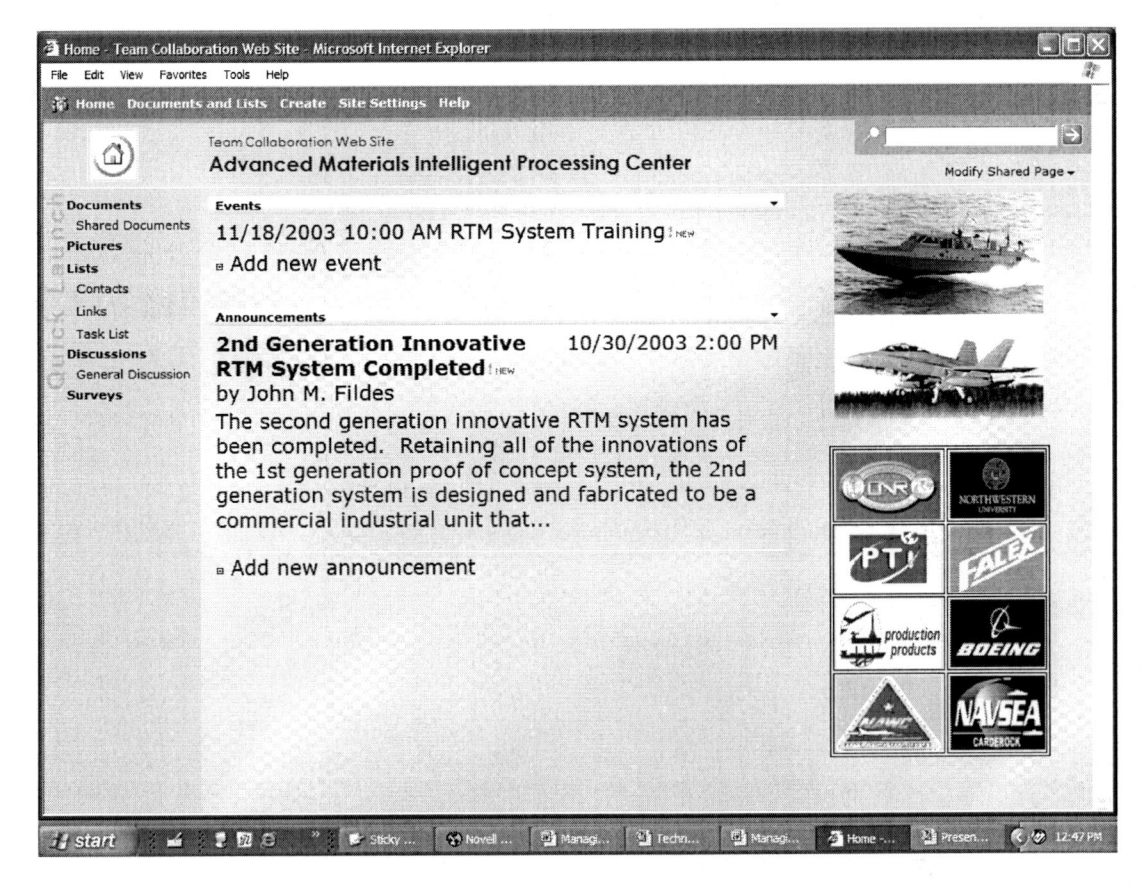

Figure 7: Design Collaboration Web Site

Figure 8 shows a document library. This library contains technical reports, presentations, engineering drawings, results of analyses, monthly status reports, and each organization's proposal. Having all documents available in one place is a tremendous asset for the team and for management of the program. For example, PTI has a library that holds all of the project milestone charts allowing the project leaders and senior management easy access to the current version.

MANAGING TECHNOLOGY
DEPENDENT OPERATIONS

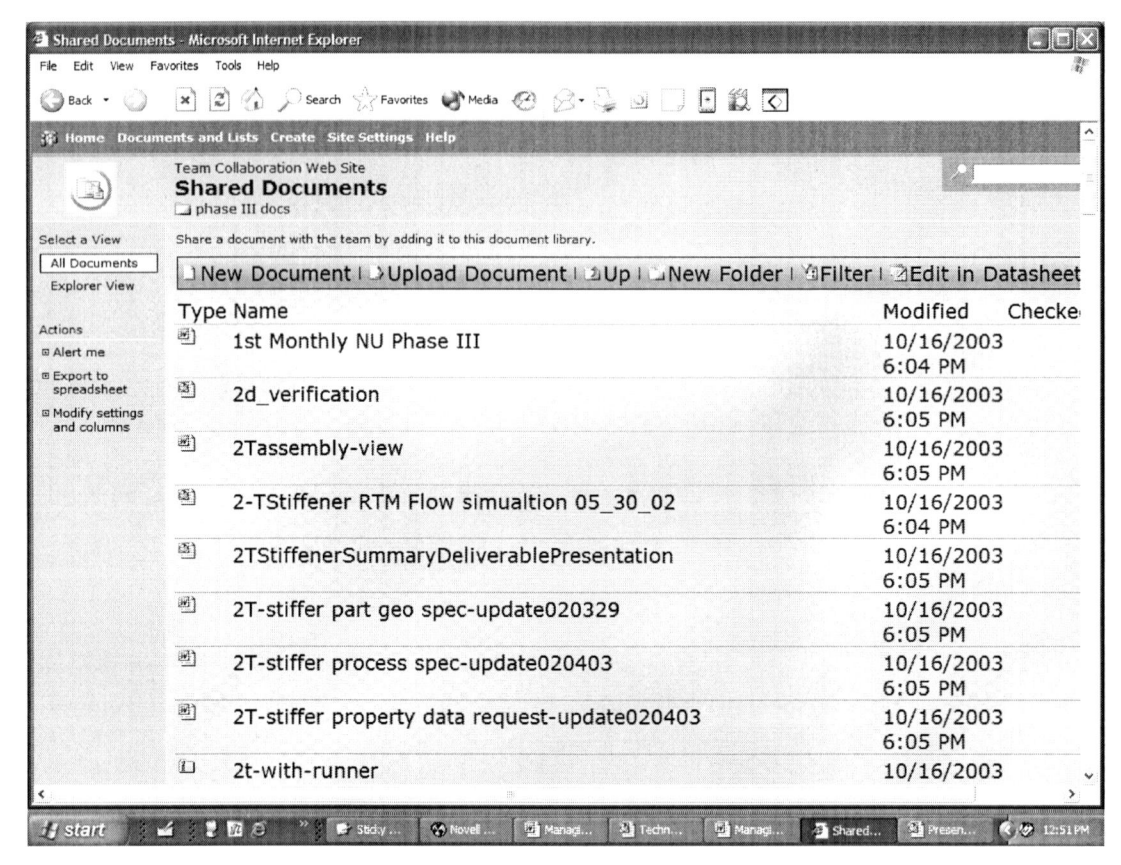

Figure 8: On-Line Document Library

The view of the document library can be easily modified. Multiple views can be defined to suite different needs. Documents in this library have been organized by the type of document and the program task to which they apply. The library can be searched and documents can be filtered and sorted so that it is easy to find documents. The name of the creator, the date of creation, the name of the last user to modify a document, and the date of the last modification is automatically tracked by the system. Each user can specify if and when they will be notified by e-mail of changes in the library.[9]

---

[9] Notification based on changes to specific documents is not currently supported by Share Point Team Services, but would be a valuable feature.

MANAGING INNOVATION
By John M. Fildes, Ph.D.

What can be done with a document depends on the type of document. Figure 9 shows an on-line task schedule. It features and uses are similar to that described for document libraries. Each user checks the task schedule for the work assigned to them and they update the % completed. This gives the customer a much better ability to monitor the status of the project.

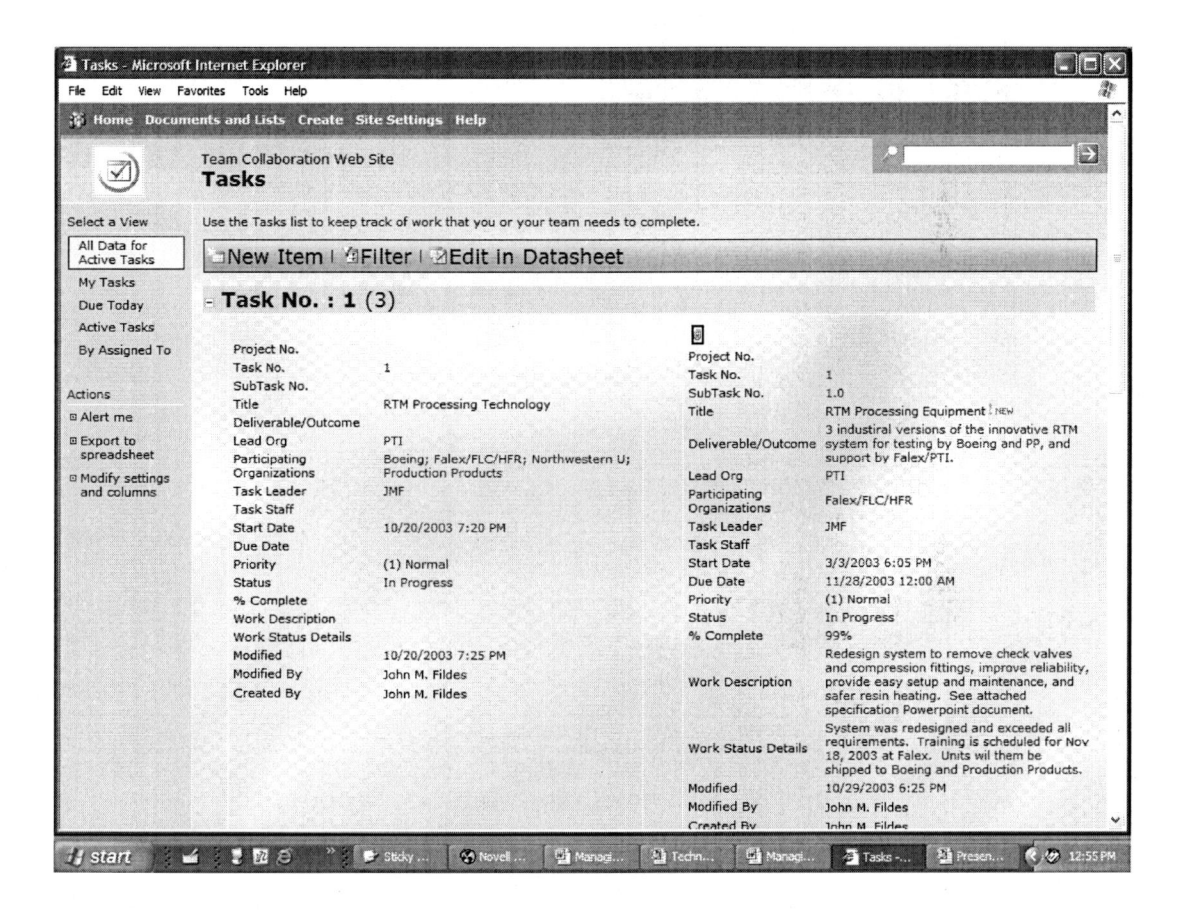

Figure 9: On-Line Task Schedule

MANAGING TECHNOLOGY
DEPENDENT OPERATIONS

Figure 10 shows a typical form that is used to enter content. Other forms are used for defining views of documents and lists. Share Point Team Services creates what is called a forms-based website. The user configures the site and enters content by interacting with a database that drives the site. This greatly simplifies the user's job in configuring and using the site.

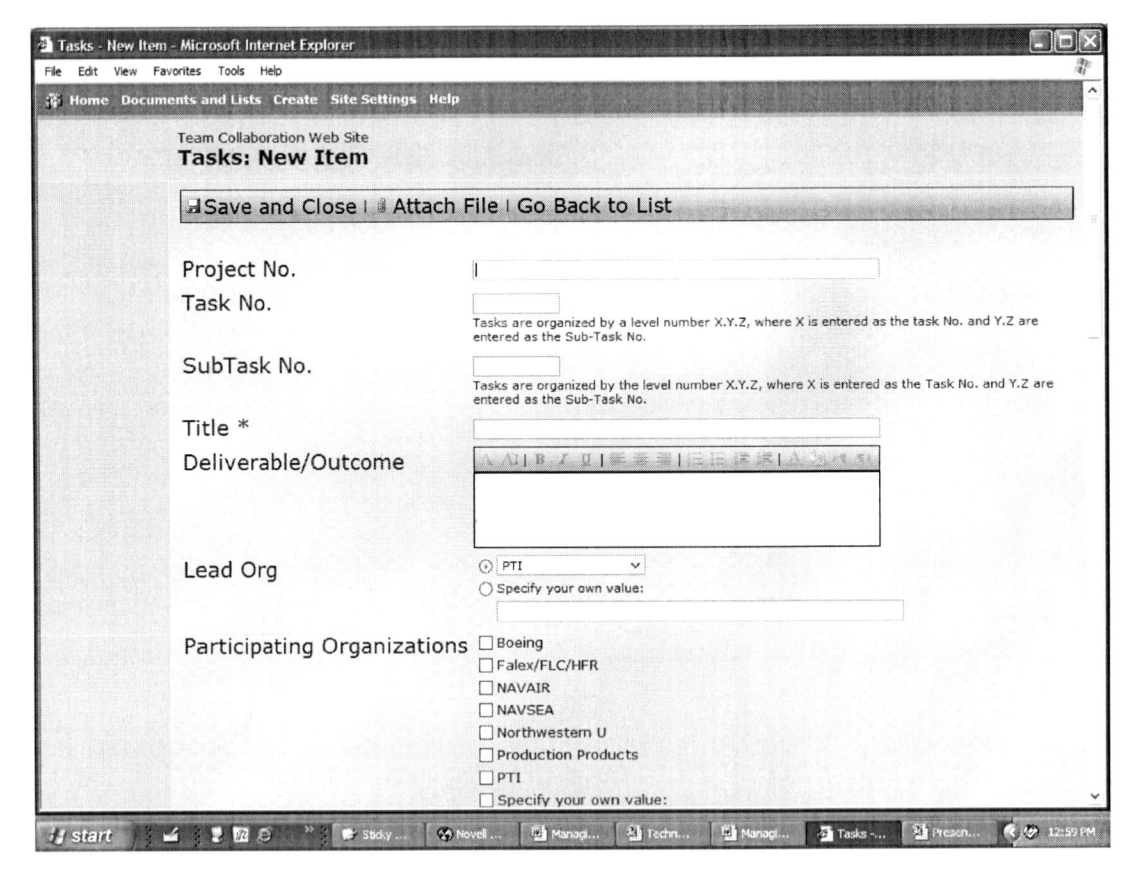

Figure 10: Forms-Based Site Management

Use of a collaboration web site offers an opportunity to employ new project coordination and management procedures, and this requires some thought and experimentation. For example in the Composites Intelligent Processing Center program, internet-based meetings are held using Microsoft Netmeeting.

Another advancement is that integrated (across team members) presentations are now possible and are used. A presentation template is placed in the document library prior to the meeting. The template contains the tasks with their objectives and lead organizations and a place to summarize the work completed in the reporting period. Each organization enters the work they completed and adds additional slides to provide backup material. Each organization always sees the full and current version of the presentation. The team used to have each organization give their presentation, leading to much duplication of background material such as objectives. The team's interactivity was also not well portrayed by these individual presentations. Now, a single integrated presentation cuts out the duplication and leads to a far more efficient meeting. The portrayal of the team's interactivity has become the central theme of the presentation.

During the Netmeeting session, one team member runs the presentation. The presenting member takes control of the presentation when it is time for their material. The interaction of the team during the on-line presentation is every bit as good as when they were in the same room. The team held quarterly review meetings at a single location for the first two years of the program. Usually, at least two participants of each organization attended. The total attendance was typically more than a dozen people. With the on-line collaboration system, the team now holds on-line meetings on a monthly basis and the meetings are more efficient. This has led to better program management and closer interaction of the team.

Netmeeting is also used for on-line engineering collaboration. 3-D views of a design can be shown by running any CAD system that runs under the Windows operating system. ProEngineer and SolidWorks are used in the Composites Intelligent Processing Center program. During the session, any participant can take control of the program as if they were actually running it. PTI routinely uses this approach to hold design development, design review, and design acceptance meetings with many of its customers. This not only saves money, it cuts time from the design process because much dead time occurs in projects waiting for a meeting so the next activity can start.

MANAGING TECHNOLOGY
DEPENDENT OPERATIONS

Figure 11 shows the redesign and conversion from a fabrication to a casting for an I stiffener made by the resin transfer molding (RTM) process. RTM is an advanced composites process that is used in the aerospace and automotive industries to make light weight parts with great strength. RTM is an automated process in which a dry fiber perform, usually graphite or glass, is placed in a closed mold and resin, usually an epoxy, is injected under pressure at an elevated temperature. Design of the part, tool, and process is technically demanding.

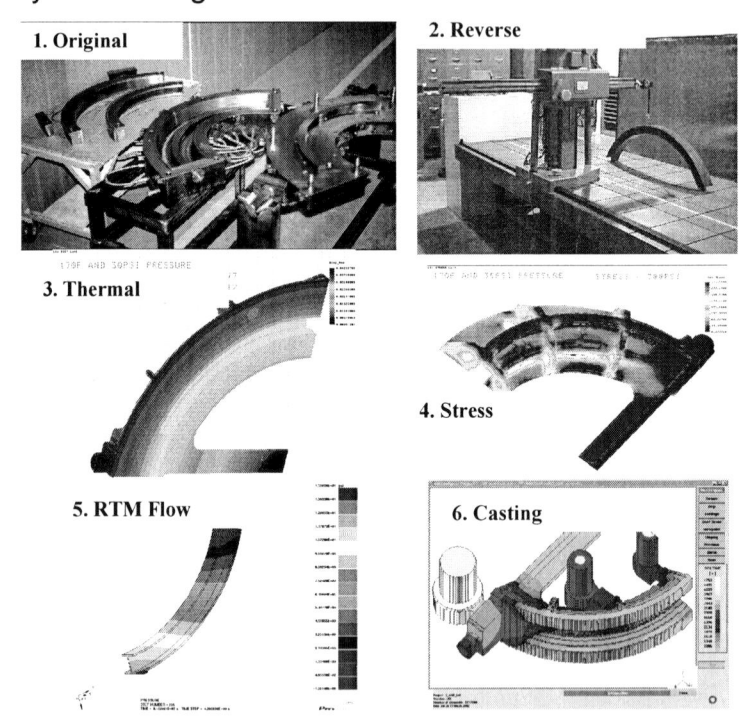

Figure 11: RTM I Stiffener Tool Designed With Innovative Process Described Herein

The tool shown in Figure 11 was originally fabricated by machining, which led to a complex, heavy, difficult to handle, and difficult to use tool. Pinching of the preform during closing of the tool, which is impossible to detect until the entire production cycle is complete, occurred in about 40%

of the runs. The preform is over $10,000 and the resin is also costly, so a 40% scrap rate had a tremendous impact on cost.

The tool was redesigned as a casting. Castings have been problematic as tools because of porosity. Extensive use of flow modeling, temperature distribution, and solidification of the casting process overcame this problem. Use of a casting reduced the mass and complexity of the tool. Additionally, heating channels can easily be included. Other innovations in the design overcame the problem with pinching of the preform. Design of the RTM process was an integral part of the tool design process, Figure 12. Flow modeling of the resin in the anistropic perform was used to improve the location of the gates and vents to achieve a shorter and more consistent infiltration of the resin. Integrated tool and process design also improve control of dimensional tolerances.

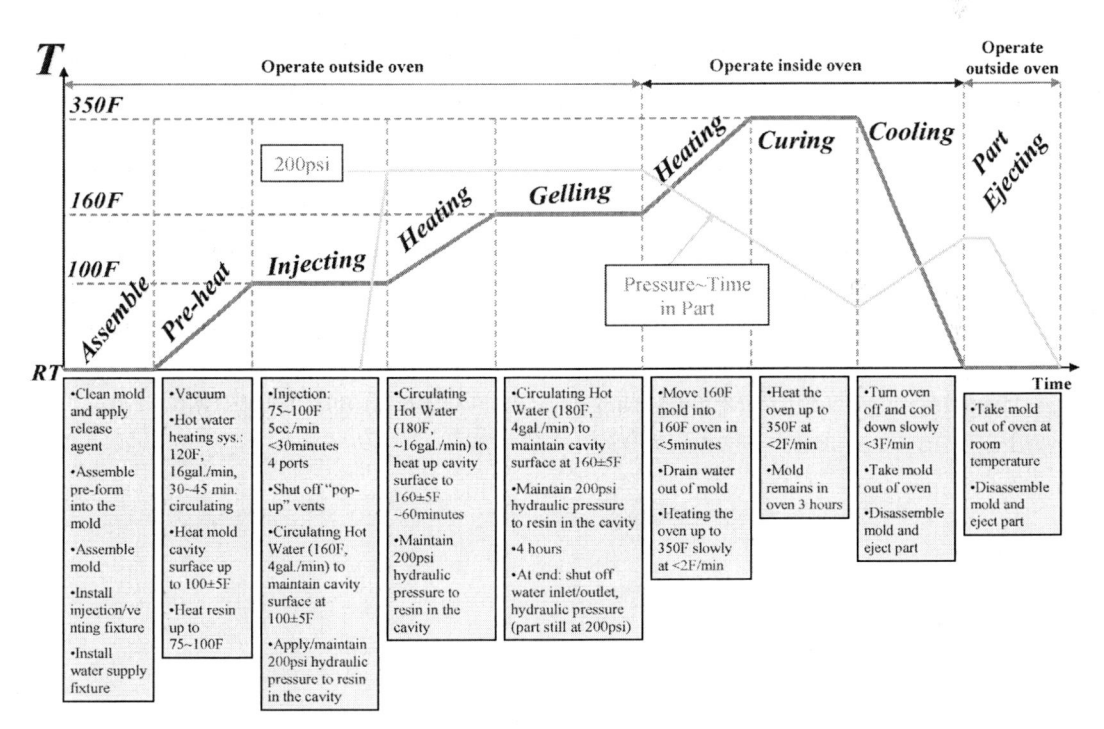

Figure 12: Integrated RTM Process Design for I Stiffener

MANAGING TECHNOLOGY DEPENDENT OPERATIONS

Figure 13 shows the process to design and validate an innovative vibration mount. This was a very challenging design problem with demands for geometry, dimensions, and the resonate frequency of the structure. This structure could probably have not been designed with conventional techniques.

The conventional design of RTM parts uses little process simulation. Also, the production mold has to be made to produce even a prototype. The investment in the production mold is so great that there is great pressure to "tweak" the mold to make it work rather than scrapping it. The design issues in the vibration mount were so demanding, coupled with the lack of accurate analytic models required an original design and two substantial revisions to achieve the desired result. This would not have been possible with conventional design techniques.

The design challenges were overcome through the creation of an innovative rapid prototyping process for the RTM mold and the extensive use of flow modeling and internet-based design collaboration. The rapid prototyping process allowed an RTM mold to be made from a solid model. Internet-based design collaboration allowed revision of the design in a matter of hours. Flow modeling ensured that the mold would perform as desired. The combined result was that a redesign was performed and a prototype mold was delivered in less than one week. The molds produced the parts as expected, allowing the bulk of the time and focus to be on testing and understanding the physics of how to achieve the desired resonate frequency within the geometry and dimensional targets.

MANAGING INNOVATION
By John M. Fildes, Ph.D.

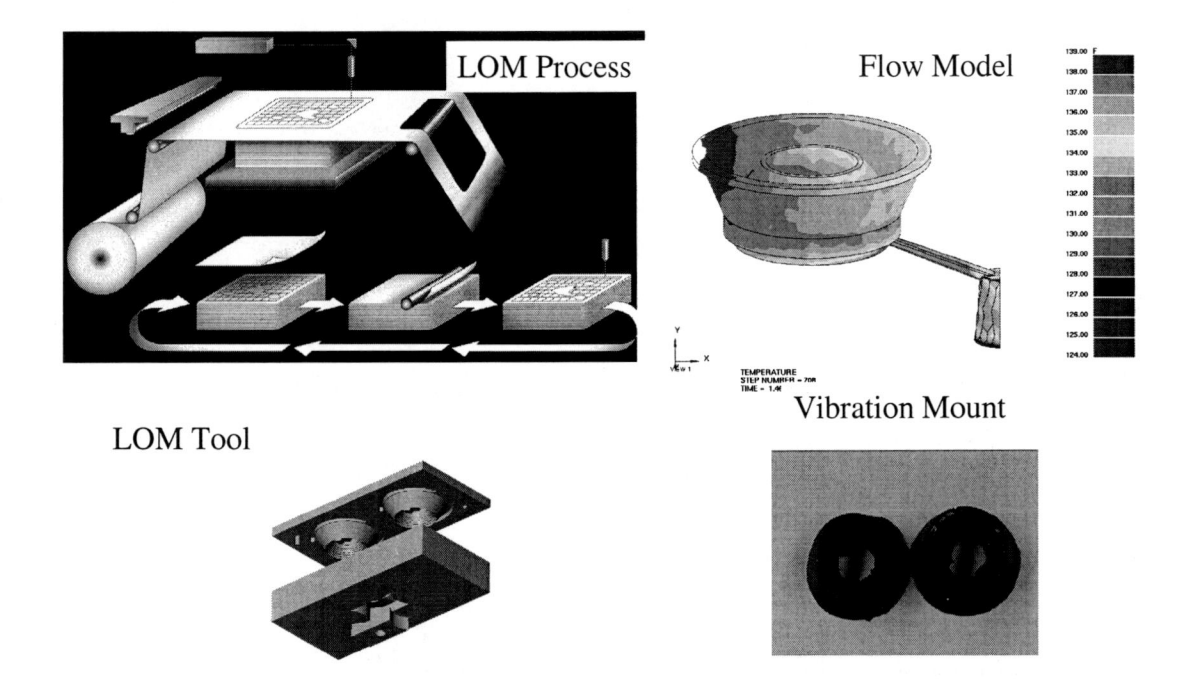

Figure 13: Innovative LOM RTM Prototype Mold

MANAGING TECHNOLOGY
DEPENDENT OPERATIONS

Summary

Innovation needs to be distinguished from creativity and invention. Innovation is a process in which a central feature is transferring an idea into action. This is true for all areas in which innovation can be manifested, be it business models, strategies, or product designs. Innovation needs to be treated as a process to be effective. This does not stifle creativity; rather it aids it by allowing ideas to become reality, avoiding the skepticism that results from not acting on creative input.

Establishing a process for innovation requires that the roles of people be clearly defined, which requires an understanding of the outlooks of various groups of people who must work together as a team. The objective and expectations must also be clearly defined. A process map is also required. The map outlines the logical set of steps that must be performed to bring about a concrete result. This approach bring consistency and allows every team to use the best practices in each situation.

Working in this environment and managing it poses demands that most scientists and engineers and many business people have not been trained to meet. Companies need to provide this training in areas such as understanding organizational structures, working with differing company cultures, virtual enterprise techniques to establish business functions that transcend internal systems and tools, and internet-based collaboration techniques that bridge geographical separation. These approaches and tools are not widely used, but they have been very successful in enough situations to conclude that they are an essential ingredient to be competitive in the future.

Chapter 6, Lead-in:

COMMUNICATIONS – MAKING EVERYONE UNDERSTAND

Neither team effort nor encouragement for creativity will be effective without good communications. How people express themselves and how others receive their messages depends on learning, habits, and circumstances.

Business managers in sales and marketing have a way of communicating that gets "a foot in the door" quickly. They look for the entry and worry about the exit later. In many cases the scientists, engineers, and technicians view a challenge from a process point of view and the validity of the outcome. They may ignore the "selling" needed to make the initial entry. Of course, it is a gross generalization, but it should be fair to say that different disciplines, different professions, develop unique communications habits.

In operations, especially those in which there is considerable dependence on technology, much of the staff is trained on making things work rather than influencing people. Nevertheless, more and more they have to influence operations strategy for they are likely to be the ones most capable of forecasting technological impact. They have to be able to "sell" their views and make effective oral presentations. This chapter should have direct value to the technical staff. Indirectly, the non-technical manager may gain a better insight on how his or her messages are received and how to interpret messages and technical reports from their staff - where to seek improvement.

Thus this chapter, while it touches on the differences in communications, it focuses more on the basic tips and tools to make communications more effective with the hope to help the organization as a whole.

MANAGING TECHNOLOGY
DEPENDENT OPERATIONS

Chapter 6

COMMUNICATIONS – MAKING EVERYONE UNDERSTAND

How to Communicate Technical Information within the Organization

Michal Safar

Introduction

Good communication skills are essential for a successful manager. A manager depends on the performance of his or her staff to accomplish departmental objectives. As the team leader, the manager needs to be able to communicate these objectives. Conversely, he or she needs to understand from their staff what resources are needed to meet the objectives. None of this happens in a vacuum, but requires a steady stream of information flowing back and forth. To the extent that everyone understands and acts on the information, the department will be successful.

Management is not an exact science nor is the selection of managers. For this reason, managers have widely varying backgrounds and skills. Those who fail, usually do so through lack of the ability to communicate clearly. Technical organizations provide special challenges for the communications process in that the information that drives product development, engineering, and production is in formats that are not accessible to the lay person, or even to people in different technical areas. Technical people have to explain sophisticated technology so non-technical people can understand. Non-technical people must also be prepared to learn some technology - it is a two way street.

This chapter is designed for use by managers of technical areas. This manager can be an engineer managing engineers or a non-technical professional managing technical staff. However, this chapter is also for those in technical environments who are not yet managers. The basic assumption is that the fundamentals of good communication must be learned by all managers and are valuable workplace skills for everyone. The emphasis here will be on the specific challenges created by the communication of technical information.

This chapter specifically focuses on communicating technical information within the organization. As such it does not address the wide range of communications to the outside, such as promotional materials, presentations at conferences, web site development, and the host of communication processes that are directed outwards from the enterprise. What it does address is the variety of communications that the technical manager must make on a daily, weekly, or regular basis that include technical information.

A lot of this chapter will be devoted to written communications; however the chapter aims to cover other forms of communication also. In the end, most communication is accomplished through words – either the written word or the spoken word. Diagrams, photographs and other pictorial visualizations are seldom successful as stand alone communication. That is why silent movies used subtitles. Therefore, this chapter will focus on organizing and presenting words effectively.

This chapter will help the manager to:
- Understand what technical information is and where to locate it
- Develop skills for organizing technical information
- Learn how to communicate clearly and effectively

MANAGING TECHNOLOGY
DEPENDENT OPERATIONS

- Prepare written and oral material in specific formats
- Develop and refine interactive communication skills
- Develop skills for understanding technical information

In summary, this chapter will focus on internal workplace communications that involve the communication of technical information. It is not meant to be comprehensive, but rather a quick reference for technical managers. At the end of the chapter is a concise listing of some basic material for additional reading that includes in depth coverage of the topics presented here and broader coverage of communications in general.

How Technical Data Becomes Technical Information

Technology based organizations accumulate technical *data*. Raw data is not easily understood. To communicate data it has to be presented in an organized format and, depending on the audience, explained or interpreted. Data that has been organized and explained becomes useable *information*. Successful communication of information results in *knowledge*.

> *...communication of information results in knowledge.*

A manager needs to be familiar with the different types of technical data, including how to read, analyze, interpret and explain them. A manager is responsible for making decisions and recommendations. These will be based on technical data and will affect the performance of the department. He or she needs to justify these decisions and recommendations. Technical information will provide the justification.

A sampling of the types of technical data an organization might have includes:

- Statistics
- Drawings
- Specifications
- Test Results

COMMUNICATIONS – MAKING EVERYONE UNDERSTAND
By Michal Safar

- Instructions/Manuals
- Lab Notes
- Analyses
- Schedules and Milestone/Charts

This is the type of data that technology based organizations generate and that technical managers need to communicate as understandable information. Some of the ways technical data is organized into information include:

- Books
- Journals
- Technical Reports
- Conference Proceedings

In order to manage and communicate technical information, the technical manager needs to know where to find the data and information within the organization. Technical information is the lifeblood of the organization. It feeds product development and production. It is important for the manager in a technical organization to know where to locate both the raw data and the distilled information, and it is his or her responsibility to educate themselves on its whereabouts and methods of accessing it. Most organizations have technical data and information in different locations depending on the whether it is raw data or information, digital or hard copy. Locations include:

- Libraries – Hard copy and digital technical information
- Internal Web Sites – Digital technical data and information
- Shared drives on Computer Networks – Digital technical data
- Technical Archives – Hard copy technical data
- Internet – Digital technical data and information

MANAGING TECHNOLOGY
DEPENDENT OPERATIONS

Many organizations have digitized their technical data. These data will reside on internal web sites, computer networks, knowledge management systems, and any location where electronic data is stored and accessed. These systems are designed for easy access with minimal instruction. They can be scattered throughout the organization, so it is a good idea to make an effort to identify all sources.

Older organizations will also have a hard copy archive of material that was not deemed worth the expense of digitizing. Access to this information is usually difficult, however it is well to be aware of its existence.

Many organizations engaged in technical research and development have special libraries that focus on technical information. These libraries will typically have the technical information generated by the organization including technical reports. They will also have books, journals, reference materials, and other technical information collected in relevant technical areas. This information will be either hard copy or digital depending on cost and ease of access.

> *The Internet has vast technical resources.*

Finally there is the Internet. The Internet has vast technical resources. Accessing information on the Internet could be, and is, the subject of full length books. A detailed discussion of technical information on the Internet is outside the scope of this chapter. The following section provides a very brief introduction to locating technical information on the Internet.

Nearly every conceivable type of technical information is available on the Internet. It is a major resource for technical information in a variety of areas ranging from standards and specifications, to government statistics, to trade publications, to patents. Because of its size, finding information on the Internet can be a real challenge. Experience and knowledge of specific technical areas are not enough to successfully mine the Internet. Fortunately there are widely available search engines that make the challenge more manageable.

COMMUNICATIONS – MAKING EVERYONE UNDERSTAND
By Michal Safar

Internet search engines provide nearly instant access to information on the web. Some of the major search engines include Google, Alltheweb, Yahoo, and MSN Search. Over the years, Internet search engines have become more sophisticated and easier to use.

They have established standardized search methods that are based on Boolean logic and employ easy to understand search formats. Most search engines provide guidelines for their use, and it is helpful to review these instructions before using an individual search engine. The following is a very brief guide to the standard search principles.

Boolean operators form the foundation of Internet searching and can be represented in various formats. Understanding Boolean logic is not necessary to do Internet searching, but it can significantly improve search results. The principal Boolean operators are **AND**, **OR,** and **NOT**. Most search engines handle these operators through search screens, so it is usually not necessary to enter them. It is, however, useful to understand how they operate and affect search results.

The Boolean **AND** is exclusive, that is it limits the search. It can be abbreviated in a search to + (plus). Many search engines default to the Boolean AND operator. Therefore the search:

dogs AND cats   or   *+dogs +cats*   or   *dogs cats*

will display web sites that have information on both subjects. This is sometimes referred to as "all of the words."

The Boolean **NOT** is also exclusive. It can be abbreviated in a search to – (minus). The search:

dogs NOT cats   or   *+dogs-cats*

will display only web sites that are about dogs and have no references to cats. Search engines refer to this as "none of the words."

MANAGING TECHNOLOGY
DEPENDENT OPERATIONS

The Boolean **OR** is inclusive. It widens the search. The search:

dogs OR cats

will display all web sites about either topic. Search engines refer to this as "any of the words." The Boolean **OR** is generally not used in Internet searching because it is very broad and generates so many search results. It does not work in some search engines.

One other easy-to-use Internet search aid is the proximity search. This is done by enclosing a specific phrase in quotation marks. Not only must all of the words be in the search results, but they must be next to each other and in the same order. The search:

"coordinate measuring machines"

will identify web sites that cover that specific topic, not just web sites on machines or measurement.

In this section the types of technical data and information have been introduced along with methods for locating the information within an organization or on the Internet. Now it is time to learn how to communicate the information.

Steps in the Communications Process – Organizing

Fundamental to the communications process is good organization. The individual who can organize his or her thoughts and information and present it clearly will successfully communicate. Below are six simple steps for organization:

1. Define the Purpose. What is the desired outcome of the communication? Most workplace communication is intended to inform, persuade, recommend, teach, or facilitate. To get started, the manager needs to write a clear purpose statement that includes the intent, the approach, and the outcome:

"This proposal will present an <u>evaluation</u> of two coordinate measuring machines and <u>recommend</u> the best for <u>installation</u> in the quality department of Titanium Products, Inc."

Intent – *Recommend*   Approach – *Evaluate*   Outcome – *Installation*

2. <u>Identify the Audience.</u> Who is the information directed towards? This will determine the style, content, and level of technical detail. With workplace communications there are several considerations. How knowledgeable is the audience? What is his/her/their position in the organizational hierarchy? Is the audience an individual or group? The focus here is on in-house communication – manager to employee, manager to group, and manager to management.

> *Who is the information directed towards?*

Managers in technical departments are required to be good at communicating at all levels within the organization and need to be aware of differences in approach for specific audiences. Communicating technical information to a technical expert requires minimal background and explanation. Sometimes it is acceptable to exchange raw data. Communicating technical information to a non-expert requires background, organization, and explanation.

3. <u>Select the Approach.</u> The means of communication is usually self-evident once the purpose is clearly defined and the audience identified. The following list matches the purpose with format:

Inform – Technical Reports, Resumes, Web Pages
Persuade – Proposals, Brochures
Recommend – Evaluations, Proposals, Presentations

MANAGING TECHNOLOGY
DEPENDENT OPERATIONS

Teach – Manuals, Instructions, Presentations
Facilitate – Memos, E-Mail, Group Meetings, Telephone calls

Because the focus here is on technical communication in the workplace, this chapter covers the essentials needed by managers to create technical reports and proposals, conduct meetings, make oral presentations, and use e-mail effectively.

4. Collect Information. Once the purpose, audience, and approach have been settled, it's time to locate the technical information that will support the communication. As was discussed earlier, there are a number of locations within the organization where information is stored. Access to the information is not always easy, however there are usually experts within the organization who will know where the information is located. When collecting information it is important to select only what is relevant to the purpose.

5. Organize. Organization is critical to successful communication of technical information. The following is one way to organize information into a clear communication. To paraphrase the old journalism maxim:

"Tell the reader what you are going to say. Say it. Tell them what you've said."

> *The audience wants to know in advance what is coming.*

Workplace communication is not creative writing. The audience wants to know in advance what is coming and how they are expected to respond.

The first step is to create an outline. Every story has a beginning, middle and end. Start with the simplest outline:

    a) Introduction
    b) Technical Content
    c) Conclusion

COMMUNICATIONS – MAKING EVERYONE UNDERSTAND
By Michal Safar

The next step is to review the material and organize it into these three chunks, then gradually to add detail to the outline. For a technical report the expanded outline might include

    a) Introduction
    b) Problem Statement
    c) Technical Approach
    d) Results
    e) Analysis
    f) Conclusions
    g) Recommendations

Once the outline headings are defined, the next step is to make a list of key points under each heading, always considering the defined purpose and audience. Does it make sense? Do the topics transition understandably from one subject to another? Once the detailed outline is complete, it is time to tell the story.

6. Tell the story. Once all of the above steps have been completed, it is time to put together the words that will tell the story. The objective is defined, the audience identified, and the information collected. The rest is easy, right? Not. Communicating clearly and effectively is not complicated, but it is difficult. The following section will cover some tips and guidelines for composing communications for maximum effect.

Using Words to Build Communications

Communication is mostly about the use of *words*. Using words effectively results in clear communication. Words are building blocks. They are used to build *sentences*. The sentences are in turn used to build *paragraphs*. The purpose of composition is to organize the material to make it understandable to the reader. The objective in

> *Holding the reader's attention requires words, sentences, and paragraphs that are clear*

composition is to hold the reader's attention. This requires words, sentences, and paragraphs that are clear in meaning and interesting. There are three components in the process of composing communications:

- Using words correctly and effectively
- Building effective sentences from words
- Constructing meaningful paragraphs from sentences

Using Words

> *...active verbs will give added impact*

When selecting words to use in a sentence, the writer should remember that the intent is to communicate, not to obscure. Therefore the writer should use the simplest words possible that accurately convey the thought. He or she should use specialized technical terms only when the reader will understand them. Above all, words need to be used accurately. Whenever the writer is unsure of the meaning of a word, he or she should look it up in a dictionary or use a different word.

The use of *active verbs* will give added impact. This means avoiding the verb *to be* in favor of more forceful verbs. For example:

Passive: *Each employee was given a bonus by the company.*
Active: *The company gave each employee a bonus.*

Passive: *The inoculation is a protection against re-infection*
Active: *The inoculation protects against re-infection*

Notice that the more active verb also results in the shortest sentence.

Writing Sentences

A sentence is a grammatical structure that conveys a complete thought. Short, meaningful sentences will hold the reader's attention. Sentences should be easy to understand. The simplest sentence contains a subject, verb, and object:

COMMUNICATIONS – MAKING EVERYONE UNDERSTAND
By Michal Safar

The consultant recommended changes.

- Subject – Consultant
- Verb - Recommended
- Object – Changes

This sentence is very effective. Contrast with the following:

A variety of changes were being recommended by the well-known consultant.

The second sentence adds a lot of words, but not more meaning. The objective should be to write short, simple sentences wherever possible. For example:

> *...write short, simple sentences*

Passive: *Chuck made an enumeration of his main points*
Active: *Chuck enumerated his main points*
Simple: *Chuck listed his main points*

Constructing Paragraphs

A paragraph is composed of a group of sentences expressing one central idea. A paragraph is complete in itself and is also a subdivision or part of something larger such as a technical report. The topic sentence is one sentence that introduces the subject of the paragraph. The sentences that follow the topic sentence of a paragraph will develop the central idea of the topic. It is important to remember that each sentence must deal only with the stated topic and not stray off into other topics. The last sentence of a paragraph will restate the idea expressed in the topic sentence and will transition to the next paragraph.

There are several ways to organize paragraphs. Some are listed below:

- *General to Particular* - In this type of paragraph order, the paragraph begins with a general statement and then moves to a particular application of that statement.

MANAGING TECHNOLOGY
DEPENDENT OPERATIONS

- *Particular to General* - Paragraphs of this kind begin with a particularity and then move into a general development.
- *Spatial Order* - Paragraphs of spatial order follow geographical direction or move from one place to another. North to South — East to West —up to down — across, etc.
- *Chronological Order* - This kind of paragraph starts with the first occurrence in a story or idea, and moves in the order of the events as they occur.

This section has touched very briefly on the basics of using words, sentences, and paragraphs in building communications. Creating clear and effective technical communications is not easy. It requires thought, time, effort, and practice. Taking the time and effort to master these skills will make the process of producing reports and presentations, as described in the following sections, much easier.

Structured Communications – Proposals, Progress Reports, Technical Reports, and E-Mail

Structured communications are primarily one-way. They are designed to be stand-alone and presented in a consecutive fashion. Day-to-day business requires the production of a variety of communications that include letters, memos, e-mail, resumes, technical reports, proposals, brochures, and presentations to name a few. This section covers the types of structured communications associated with the transfer of technical information. The primary focus of this section is to cover the documents associated with the life cycle of a technical project. These are the *technical proposal, project progress report,* and *technical report. E-mail* will also be covered as it is used to communicate technical information. Although e-mail is considered to be an informal means of communication, to be effective it should be well constructed.

> *Structured communications are primarily one-way.*

COMMUNICATIONS – MAKING EVERYONE UNDERSTAND
By Michal Safar

A technical organization routinely undertakes technical projects. Part of good project management practice is complete and accurate documentation of the activity. Some of the most common types of projects in technical operations include:

- Process/product research & development
- Productivity and Quality improvement
- New facilities planning, equipment acquisition & installation
- Contract research & development
- Software acquisition, implementation, & development.
- Safety and maintenance programs

Throughout the project life cycle the following reports document the technical progress: The *proposal* persuades the organization to invest in the project; the *progress report* provides periodic project status; and the *technical report* documents all of the activity and results from the entire project. All of these reports include technical information. The manager of the department is responsible for preparing, or overseeing the preparation of, this material. The better it is written and presented, the more successful the project will be. This section will cover the life cycle of a project to replace an aging coordinate measuring machine.

> ...the *proposal* persuades... the *progress report* provides status...the *technical report* documents...

Proposal

In following the preliminary steps for communicating outlined above the writer will define the purpose/objective and consider the audience. If the decision has been made to prepare a proposal, then the purpose and objective are already defined. Project proposals need to get attention of the right managers. The audience for a proposal will typically be upper management. In some cases big reports and proposals for things far away in time or location get put aside. A cover letter may be needed to get the

MANAGING TECHNOLOGY
DEPENDENT OPERATIONS

attention of the right manager. It should point out up in front of the expected outcome or clearly tell the decision maker why he or she should open the cover and read it – what can he or she expect, what is in it for him or her.

The proposal needs to convince the audience of the need for the project, convince them that it's worth the cost, and finally, convince them that there is the expertise to complete it successfully. These points need to be clearly incorporated into the proposal.

The proposal format is as follows. This section will cover each component.
- Abstract
- Introduction
- Research
- Technical Approach
- Desired Outcome
- Attachments

> *...all proposals should start with an Abstract.*

A proposal should be a highly structured document. Its purpose is to clearly convince. It is not a mystery novel where the solution comes as a surprise at the end. For that reason, all proposals should start with an Abstract. The abstract summarizes the entire document. It is the single most important paragraph in the proposal. The well-written abstract will insure that the body of the proposal receives consideration. A poorly written abstract, or the lack of an abstract, can result in the rejection of the proposal. Individuals in management have limited time. The purpose of the abstract is to convince the reader to invest the time in reading the whole proposal.

The abstract can be as short as a single paragraph but should not be longer than a page. It will mirror both the content and organization of the proposal with one sentence for each major point covered. For this reason, the abstract should be written only after the body of the proposal has been completed.

The introduction sets the stage for the proposal and covers three areas. Following the example listed earlier involving coordinate measuring machines, the introduction will start with a *background* of the use of coordinate measuring machines (CMMs) within the organization. Then the proposal will *identify the problem,* which is that the existing CMM is old, not repairable, and needs to be replaced. It might describe how current operations are adversely impacted by the existing equipment. What follows is the *purpose* of the proposal, which is how the problem will be addressed. In this example, two or more CMMs will be evaluated and a recommendation will be made about what one to purchase. Finally, the proposal will identify and define any *technical terms* that may not be familiar to the audience. In this case it might be appropriate to include a short tutorial on what a coordinate measuring machine does and identify key functions and components.

> *The introduction sets the stage...*

The next step is to research the subject, summarize the findings, and make a recommendation. In the case of CMMs, the manufacturers of the equipment would be contacted for product literature, pricing information, and specifications. Trade journals could be consulted for articles that evaluate CMMs. Next the proposal will include the evaluation of the equipment capabilities and cost against the specific requirements the department has for a CMM. Finally, the proposal will present a recommendation on which CMM to purchase. This section of the proposal is meant to present persuasive evidence to support the recommendation. As such it will be a summary of the findings with selected quotations from the material researched. This is not the place to include specifications or detailed product literature. That information is appropriately included in the Attachments section.

The technical approach details how the proposal recommendation will be implemented. This will include a project plan that describes the

> *The technical approach details the implementation.*

203

requirements for removing the old machine and installing the new machine. Any building modifications required will be identified along with possible disruptions to the production schedule. The technical approach is usually structured as a series of tasks. The tasks are scheduled according to dependencies and can be effectively displayed in a Gantt chart. Some scheduling considerations include lead time from order to delivery, installation time, testing and training time. The technical approach will also address cost. It is important to include all of the cost elements including the cost of the equipment, cost of any service contracts, cost of shipping and installation as well as the cost of in-house staff time and possible loss of production time. Finally, there will be a risk assessment. This section will identify what the risks of the project are, what might go wrong, and what the project plan does to mitigate the risk. This is where to detail the qualifications of the firm selling the equipment as well as the qualifications of the in-house staff to complete the project.

The last part of the body of the proposal is the desired outcome. This will include a schedule for follow up on the installation to insure that training, and any service contracts are completed as scheduled. It would also be appropriate to include a plan to monitor and evaluate the performance of the equipment. Finally, if possible, the proposal should establish an anticipated payback time for the equipment.

The attachments section will include product literature, specifications, and any of the other research material relevant and that the reviewers may want to see.

The more complete and well written a proposal is, the fewer questions and challenges it will face, and the faster it will be funded and implemented.

Progress Report

The next stage of documentation for the project life-cycle is the progress report. This is a periodic status report for the project that goes on for the life of the project. It should be well structured and concise. This report is meant to be a snapshot of a specific period of the project – monthly or quarterly - to assist those who are involved in the project. The recipients

COMMUNICATIONS – MAKING EVERYONE UNDERSTAND
By Michal Safar

should include those involved in oversight of the project as well as the staff who are working on the project. The key to an efficient progress report is to use as few words as possible and employ tables and charts that can be easily updated from period to period. A sample progress report format is as follows:

> *An efficient progress report uses as few words as possible.*

- Introductory Material
- Task activities
- Expenditures
- Schedule status
- Outstanding problems & issues
- Activities for the following period

The introductory section will include the project name, number (if appropriate), project manager, date and time period covered. It will have a brief paragraph describing the project.

The task activities section will include a brief description of each task, give a summary of activities performed during the period, and list the task status - underway, complete, not started, or suspended.

Project expenditures are best covered in a spreadsheet. Expenditures should be tracked against the budget for both the current period and the cumulative period. Expenditure categories should include labor, materials, overhead and travel.

The schedule is section that is best handled as a table chart as follows (Figure 1):

MANAGING TECHNOLOGY
DEPENDENT OPERATIONS

Figure 1: Shedule Chart

Task	Task Description	1	2	3	4	5	6	7	8
	Task 1	■	■	■	■	■	■	■	■
1.1	**Subtask Baseline**	■	■	■					
	Status			C					
1.2	**Subtask Baseline**	■	■						
	Status						M		
	Task 2	■	■	■	■	■	■	■	■
2.1	**Subtask Baseline**			■	■				
	Status					C			
2.2	**Subtask Baseline**				■				
	Status						C		
	Task 3			■	■	■	■	■	■
3.1	**Subtask Baseline**			■	■	■			
	Status						P		
3.2	**Subtask Baseline**						■	■	■
	Status								P
	Project Management		■	■	■	■	■	■	■
4.1	**Project Management**		■	■	■	■	■		
	Status						O		
4.2	**Management Reporting**			■	■	■	■	■	■
	Status						O		

O	On Schedule, Ongoing
P	Postponed/On Hold
M	Missed Schedule
C	Complete

This particular schedule makes it easy to see the baseline schedule for each task as well as the task status.

One section should be devoted to <u>identifying problems</u> or issues that need to be resolved. The problem description should include information about the actions taken or planned to resolve the problem and an estimated date for the solution.

The final section should address the <u>activities planned</u> for the next reporting period. This includes task activities, meetings and travel.

Technical Report

The final stage of the project life cycle is the technical report. This is produced on completion of the project and serves to document the entire project. The audience for this report is broad – anyone within the organization including upper management, scientists, researchers, engineers, and technicians. For this reason there needs to be different access points to the information in the report. Very seldom is a final technical report read from cover to cover. Most readers will read only the executive summary or a specific section or chapter of interest to them. Some will be interested only in the technical data. The two access points in a technical report are:

> *The final stage of the project life cycle is the technical report*

- *Executive summary* – This is a high level overview of the project for those who are interested in the project description and results, but are not interested in all of the technical details.
- *Table of contents* – This lists all of the sections, tables, figures and appendices in the report and provides quick access to specific information in the report.

> *The executive summary is the single most important section.*

The information in the technical report will be taken largely from the information already collected and documented in the Proposal and Progress Reports. The format of the technical report is as follows:

MANAGING TECHNOLOGY DEPENDENT OPERATIONS

- Front Matter
- Executive Summary
- Technical Discussion
- Conclusion
- Appendices

Because a technical report documents the project for posterity, it includes more detailed information and is carefully formatted. Most technical reports will include extensive *front matter*. Front matter includes a title page with the project name, personal author(s), corporate author, date, project number, and any other internal reference information. Typically the report will have a preface or foreword that makes acknowledgements and credits regarding funding, technical information, or other contributions. A detailed table of contents will include all of the section titles, section subtitles, and a listing of any figures, tables, and appendices.

Like the abstract in the proposal, the *executive summary* is the single most important section of a technical report. The executive summary is a highly distilled summary of the project history, activity, results, and conclusions. Again, this is not a mystery story. The audience wants to read the end first. Like the abstract in a proposal, the executive summary will be written after the rest of the report is complete. It is a stand-alone document and should be no more than two or three pages in length. The executive summary is frequently the only portion of the report most of the audience will read. It needs to be clear and concise.

> *The audience wants to read the end first.*

The organization of the *technical discussion* will be similar to that of the proposal. It will include an introduction that explains the history and problem statement. It will go on to describe the technical approach, usually by task or chronologically. In the technical discussion the report will document the original project plan and compare it to the project as it actually evolved. A comparison of the original schedule to the actual time line should also be included. Finally, the results of the project should be presented.

The *conclusion* section will include a summary of the project as well as conclusions. Was the project a success or failure? Were the outcomes as expected? If not, what were the differences? Finally, the report will detail the lessons learned and make recommendations. In the instance of our coordinate measuring machine installation one of the lessons learned might be, *"Future similar projects should include a longer equipment delivery lead time."*

The last section of the technical report will include any *appendices* that may provide supporting technical information that is too detailed to include in the body of the report. Material appropriate to the appendices includes detailed research results, peripheral material of relevance, references, raw data, descriptions of methodologies, equipment specifications, floor plans, and technical drawings. In a very detailed technical report an especially useful appendix is a list explaining any acronyms and abbreviations used in the report.

The preparation of the technical report concludes the life cycle of a project from inception, through implementation, to final documentation. During any technical project there is ongoing, less formal communication of technical data and communication through a variety of means. One additional written communication this chapter will cover concerns the use of e-mail to exchange technical information.

Electronic Communications – E-Mail

Electronic communications have revolutionized the exchange of all types of information, including technical information. At the individual level, e-mail is replacing the telephone and other forms of communication. E-mail allows an individual to communicate directly with multiple recipients. It is easy to use. Detailed technical documents can be sent digitally as e-mail file attachments very quickly. Unlike telephone calls or face-to-face meetings, e-mail is written documentation of a communication that can be saved for later reference.

MANAGING TECHNOLOGY DEPENDENT OPERATIONS

> *E-mail is often deleted before being read.*

There are, however, drawbacks. Because e-mail is so easy to use, most people receive more messages than they can easily handle. As a result, e-mail is often not carefully read and frequently deleted without detailed review. When communicating technical information, an incoming e-mail message is in competition with the entire contents of the recipient's in box. Another drawback is the lack of interaction. E-mail does not convey the tone of voice, urgency or other emotional nuances. E-mail needs to be read for content, not intent. This section will cover some of the techniques for composing professional level e-mail that will be read and responded to. The following are some guidelines for producing e-mail messages that communicate clearly and effectively:

Provide specific information in the subject line. Frequently, the subject line is the only portion of an e-mail message that the recipient will look at. The decision to delete or open an e-mail message is sometimes based on the content of the subject line. It is important for the sender to make sure that the subject is clear and will be understandable to the recipient.

> *E-mail messages should be composed as short paragraphs.*

Construct easily viewed messages. An e-mail message without paragraphs or line breaks is difficult to read. E-mail messages should be composed as short paragraphs with blank lines separating the paragraphs. This way the message will stand out. The use of colored text, in moderation, can help to make clear the gist of the message. However, elaborate formatting should be avoided. There are variations in e-mail software. Features supported on one e-mail viewer may turn out to be unreadable garble on another.

Keep messages short. Each message should be limited to one topic. If there are multiple topics, separate messages should be sent for each. People who have a large volume of e-mail to get through every day frequently read only the first paragraph of each message. If there is more than one topic, it will not be read. E-mail should always contain a greeting and a signature. This will focus the reader on the message. If a forwarded message includes several earlier messages, they should be summarized in the introductory comment. The use of colored text can help to highlight and organize long and complex forwarded messages.

Take time to reread and revise. Everyone has limited time to read and compose e-mail messages, but haste leads to inaccuracies, which in turn require additional messages to clarify. Although e-mail is considered a somewhat informal method of communication, it is still the written word. Bad grammar, misspellings, and poor sentence construction not only confuse the reader, but reflect badly on the sender. Professional level communications require careful composition.

Compose for public distribution. There is no privacy on the Internet. E-mail messages are routinely forwarded to multiple recipients. The sender has no control over where a message may be sent or who may see it. Critical remarks about others, jokes, and personal information have no place in professional communications.

Explain attachments. File attachments are an easy and efficient way to transmit large volumes of technical information, assuming that the recipient will open the attachment. Opening and reading a file attachment requires time and effort. If there is no explanation of an attachment in the e-mail message the attachment is frequently ignored. The sender should always identify the attachment in the body of the e-mail and explain its intent. A short excerpt from the attachment will help to make the content clear. The sender should always try to make the recipient understand why it is important to read the attachment.

> *...always identify the attachment in the body of the e-mail*

MANAGING TECHNOLOGY
DEPENDENT OPERATIONS

E-mail can be effectively used to communicate technical information rapidly, completely, and accurately. However, incoming e-mail messages are always in competition for the time and attention of the recipient. The clearer and more concise the message is, the more likely it is to be read.

Structured Communications with Some Interaction – Presentations

Most people dread it - the oral presentation in front of an audience. The discussion that follows here will give some tips about getting through what for many is a very challenging experience. These steps and suggestions will not guarantee the most charismatic presentation, but it will get anyone through the process and accomplish the objective, which is communication.

> *Most people dread it - the oral presentation.*

The process of putting together the information for a presentation is the same process outlined above in Steps in Communicating and Using Words. However, a presentation has some special considerations that this section will focus on. A presentation happens in real time with time constraints. The presenter usually has a fixed amount of time to deliver the message. This section will focus on the presentation preparation as it differs from preparing written documents - the delivery.

There are a variety of presentations a technical manager may be called upon to make within the organization. The way a presentation is organized will depend on its purpose. In line with the coverage of internal communications, this section will focus on presentations of proposals, project reviews, and tutorials. After an overview of the preparation and delivery specific to each of these types, the section will focus on the details of presentation preparation and delivery.

Frequently the proposal review process will include an oral presentation of the proposal. Because a written proposal already exists, the *purpose, organization and composition* of the presentation are predetermined. However, its delivery is not. There are critical differences between how a proposal should be written and how it should be presented to an audience.

In a **proposal presentation** the audience will typically be a small group, usually the decision-maker, managers from departments that may be impacted by the proposal, and sometimes representatives from financial areas. It is important to identify the decision-maker within the group and focus the presentation on that individual. The purpose here is to *persuade*, and in preparing the presentation the proposer needs to anticipate any concerns or issues that individuals in the group may have. To foster a positive, cooperative atmosphere, it is important to introduce each member of the group and identify his or her role in the organization and relation to the proposal. The presenter should express appreciation to the audience for the time they are spending to review the proposal and solicit their cooperation in its acceptance and implementation.

> *...identify the decision-maker within the group and focus on that individual.*

Timing is important in any presentation, and the length of the presentation will depend on the complexity and scope of the proposal. However, the proposal exists as a written document for the group to review on their own time. The proposal presentation should be a brief, succinct overview that covers the main points of the proposal. The proposer should never read directly from the proposal, or the audience will get lost in the detail and miss the main points. An optimum time for getting the essentials across while holding the audience's interest is 20-30 minutes, with a designated time for questions.

> *The proposal presentation should be a brief, succinct overview.*

Workplace presentations today almost always include the use of presentation software, with charts prepared to highlight the key points of the presentation. In this instance, hard copy handouts of the charts should be provided to the audience to enable them to easily follow the presentation.

MANAGING TECHNOLOGY
DEPENDENT OPERATIONS

If the proposal has not previously been made available to the recipients, it should not be distributed until after the presentation. If the audience is trying to read the proposal during the presentation they will miss the key points being presented.

A program/project review will be structured similar to the progress report. Its purpose is to *inform* peers and management about the status of a project. The audience will be people directly involved in or responsible for the program and as such will be reasonably knowledgeable about it. The time for a review varies depending on the complexity of the project, but can include a working session after the presentation to address issues and problems in the project. Charts or other presentation aids should be used to highlight the project status. Flip charts or white boards should be available if the review includes a working session.

Technical managers are frequently called on to present tutorials and seminars in their areas of expertise for the education of individuals within their departments or individuals in other departments. The purpose of such seminars is to *teach* the audience, which will usually be larger and more diverse than the previous two examples. The length of time for these presentations varies, but should be announced to the audience in advance. The presentation should be completed within the announced time. Adequate time should be allowed for questions and discussion. In addition to presentation charts, there should be detailed handouts of both the presentation and any related information.

Presentation Preparation
Presentation preparation requires in-depth audience evaluation. The audience for a written document is faceless and nameless. The audience for a presentation is alive and present. They have names and faces. They ask questions. They get bored.

Presentations require in-depth audience evaluation.

An effective presentation will be closely tailored to the audience. Some of

COMMUNICATIONS – MAKING EVERYONE UNDERSTAND
By Michal Safar

the questions that need to be asked include: How large is the audience? The size of the group will determine what presentation aids are the most effective. How knowledgeable are they about the subject? This will provide an understanding of how much technical detail they will understand. What are they expecting? If the audience is expecting a 20 minute presentation, it would be a mistake to prepare a two hour speech. Who is the decision maker? The presentation may need to be tailored to that individual. What actions are required of the audience? If their input on a topic is needed, that needs to be made clear at the outset.

The level of technical detail to include in a presentation is dependent on the background and knowledge of the audience. The speaker should never underestimate their intelligence. Nobody likes to be talked down to. On the other hand, the speaker should not overestimate their knowledge of the technical of the details of the material. He or she should be prepared to explain the technical details simply and avoid the use of "insider" jargon. However it is acceptable to use generally understood technical terms if the audience has the technical background.

> *...avoid the use of "insider" jargon*

Important considerations for both the audience and presenter are the time of day that presentation is made and its length. Morning presentations typically get the most undivided audience attention. In the late morning the audience is beginning to focus on lunch. Directly after lunch the audience has to be refocused. Late in the day the audience is thinking of going home. Mid-morning and mid-afternoon are the times when the audience is most alert and focused and most likely to be receptive to a presentation. If the presentation is scheduled for a less than optimum time, the speaker needs to be prepared for a less than attentive audience. Also, presentations are normally allowed a fixed amount of time. If a presentation exceeds the allotted time, the audience will lose interest. Presentation preparation includes allocating the appropriate amount of time for each section of the presentation.

There are other considerations when preparing a presentation. One

MANAGING TECHNOLOGY
DEPENDENT OPERATIONS

is the size of the room. If the room is large and the speaker does not have a strong voice, a microphone will be necessary. What equipment is available? If the presentation is designed for computerized projection it will be impossible to display if the meeting room is equipped only with on overhead projector.

Organizing the presentation is a critical step and as with any of the other communications previously discussed should be based on the *Introduction, Technical Discussion, Conclusion* format. If the presentation is based on a proposal or other source document, it will follow the organization of the source document. However, it cannot be overemphasized that the speaker should never read from a source document during a presentation. The objective is to identify and present the main points, not all of the detail. If there is no source document, the *Steps in the Communications Process* should be followed in organizing and preparing the presentation material.

> *...never read from a source document.*

Some specific elements should be included in the different sections of the presentation. The *introduction* should include an "ice-breaker" whenever possible. This is a short joke or story to introduce the topic and get the audience's attention. Following the opening, the speaker should give the audience an outline of the presentation and let the audience know approximately how long the presentation is. The outline should be simple, and repetition should be used to make sure the audience understands. The introduction should conclude with a phrase similar to "The three concepts I will present today are..."

The organization of the *body* of the presentation depends on the subject. As was mentioned before, if the presentation is based on another document, its organization is predetermined. In an original presentation, the organization will depend on the material. Of the several ways to organize material one is chronologically. If this is appropriate to the topic, it should be used. It is easily understood by the audience. If the presentation is

> *... repetition is important in a presentation*

highly technical the material can be presented by level of difficulty, starting with the easiest to understand material and progressing to the more difficult concepts. Another approach is hierarchical, starting with the least important and concluding with the most important topics. Whatever organizational approach is used, it should be made clear to the audience in advance.

In *concluding* a presentation the main points should be summarized. Because it is harder for people to remember the spoken word, repetition is important in a presentation. Phrases such as, "The three ideas I want you to take away with you are..." will emphasize the main points.

The final preparations for a presentation include planning the presentation aids. These include visuals, the equipment needed to display them, and handout materials. Whatever can go wrong will go wrong in front of an audience. All presentation material should be prepared in advance and all equipment tested.

Presentation Delivery

For most people the actual delivery of a presentation is a significant challenge. Presentations on technical subjects are not speeches, read word-for-word from a script. The technical presentations under discussion here assume a certain amount of interaction with the audience. The speaker must have the ability to adapt the presentation "on-the-fly" to address questions. This section covers guidelines for making an effective delivery.

> *... have the ability to adapt the presentation "on-the-fly"*

Practice. Practicing is the secret to successful presentations. The more comfortable the speaker is with the material and the act of delivering it, the more effective the presentation will be. Ways to practice include:

- Practicing out loud in front of a mirror
- Tape recording or video taping the presentation
- Giving the presentation to a friend or relative

MANAGING TECHNOLOGY
DEPENDENT OPERATIONS

An important aspect of practicing a presentation is timing it to insure that it does not run over the allowed time. There is nothing more disconcerting than to be in the middle of a presentation and be told that time is up. When this happens, the speaker tries to rush through the remaining material, and the presentation has lost its impact.

Consider appearance. Some of the elements that affect the presentation are voice, manner, and body language. A conversational tone will put the audience at ease. The presenter should speak clearly without mumbling and try to convey a manner that is relaxed, sincere, professional, and knowledgeable. His or her dress and clothing should be appropriate to the situation. Good posture will help convey a professional manner. Finally, it is important to communicate enthusiasm. If the speaker does not appear to be interested or involved in the subject, the audience will not be either.

Interact with the audience. Although a presentation is prepared in advance and structured, a successful presentation is also interactive with the audience. The speaker should prepare for interruptions and questions and respond gracefully when they occur. He or she should always look at the audience and make frequent eye-contact. This will draw the audience into the presentation.

> *...look at the audience*

Be comfortable with the presentation environment. The speaker needs to check to make sure all of the equipment needed is available and functioning. He or she should also check the stage or presentation area in advance to determine the most effective place to stand.

Accept nervousness. It is normal for anyone presenting to have some level of nervousness. With practice and experience a successful speaker can channel nervousness into energy and enthusiasm.

Presentation Aids

There are a variety of presentation aids. They include flip charts, transparencies, computer projections, slides, movies and handouts. The advantages and disadvantages of using each are covered briefly in the following section.

Flip charts are one of the oldest and most reliable presentation aids. They have no moving parts to break, no software to crash, no light bulbs to blow. They allow for spontaneous demonstrations and capture the ideas generated during the presentation. They are most effective when used in with small groups because they cannot be seen from a distance.

Transparencies are another old standby for speakers. They are easy to produce and portable. The overhead projectors required to display transparencies are almost invariably available and are very dependable in a presentation situation. Transparencies are good for larger groups. Blank transparencies can be used for capturing ideas with the advantage that they can be seen by a larger audience. However, transparencies need to be moved on and off the projector, which can distract the speaker's attention from the audience. Loose transparencies are easily mixed up.

Computer projection systems running presentation software are probably the most commonly used presentation aids and have a number of advantages. Since the display is directly from a computer file, there is no need to produce slides or transparencies. The presentation can be revised very easily and tailored to the audience. Presentation software supports a wide range of special effects including animation, sound and video. A well prepared presentation looks professional and progresses smoothly. There are some serious drawbacks, however. Computerized presentations require more resources. In addition to the screen, a projection system, computer, and software are required. Projection systems can be very expensive to own or rent, as can laptop

> *Computerized presentations require more resources.*

MANAGING TECHNOLOGY DEPENDENT OPERATIONS

computers and software. There is a high risk of equipment or software failure or malfunction. Because of the number of different projection systems and the range of laptop computers with varying resolutions and interfaces it is essential to test all of the equipment well in advance of the presentation. If there are multiple presenters, each with a laptop, switching between computers can be time consuming and distracting.

Slides were once the standard, but are now seldom used for presentations. They have been largely supplanted by computer projections. They had the advantage of being good for large groups, but are expensive to produce and changes require the generation of new slides. Slide projectors are no longer readily available and can be unreliable.

Videos are occasionally used in presentations. The projection equipment is usually reliable and easy to operate. However, there is no flexibility in a video presentation. It starts at the beginning and goes to the end. Video tapes also have high production expense.

> *Visuals should not be crowded or overly complex.*

DVD technology represents one of the recent trends in presentation aids. This can incorporate the audio and video of tapes into a computer projection system. DVD presentations are typically very flexible and extremely portable.

Handouts are another useful presentation aid and can range from copies of the presentation slides to supporting information. They provide a convenient place for the audience to take notes and give the audience a roadmap of the presentation.

There is no doubt that visual aids enhance the effectiveness of the presentation. When using any of the above it is important that they be well designed. Visuals should not be crowded or overly complex. They should be clearly labeled and easy to understand. Visuals should be consistent with the presentation style and organization and the content should be appropriate to the presentation.

COMMUNICATIONS – MAKING EVERYONE UNDERSTAND
By Michal Safar

Presentation Design

Presentation software such as Microsoft PowerPoint™ has revolutionized presentation design and preparation. No longer does the individual making a presentation need to depend on graphics artists to generate slides. Designing a series of charts using presentation software is a relatively intuitive process resulting in reasonably sophisticated and effective presentations with moderate effort. Following are some guidelines for designing presentation materials.

> *Is the purpose to inform, teach, or persuade?*

As has been repeated, ad infinitum, the audience needs to be considered in the design of presentation charts. Are they expecting a highly detailed technical presentation or a simple overview? The speaker needs to determine in advance their preferences in presentations. Some individuals are more impressed by charts that use a lot of graphics and special effects. Some individuals prefer a simpler presentation. Is the purpose to inform, teach, or persuade? The appearance of the presentation will vary with its purpose. For the type of internal presentations discussed in this chapter – proposals, tutorials, project reviews - a simple, direct style is usually the best. Fancy graphics and imbedded animations are more appropriate to the marketing and public relations departments.

This section has covered the content and delivery of the presentation. Now it will focus on a few simple tips for designing effective slides by considering the format, style, and design.

The slide *format* should be consistent without being boring. Smooth transition from slide to slide will keep the audience focused on the message. This means standardizing the positions of key elements such as headings, titles, and graphics that repeat from page to page. It also means using transition charts as the presentation moves from topic to topic. Repetition is essential to any

> *Always start with an outline.*

successful presentation. At the beginning of the presentation the speaker should start with an outline chart and reintroduce it at the beginning of each new topic. This approach serves a couple of purposes. It reminds the audience of what has been covered and what will be covered. It also gives the audience a sense of progress through the presentation and a clear view of the conclusion.

The *style* of the charts should also be consistent. This means using one or two fonts throughout the presentation. The fonts should be readable, and the text should contrast with the background. If the background is dark, a white or light text is appropriate. If the background is white, a darker text should be used. Design elements such as graphics, video and sound clips, and complex animations should be used strictly to support the content and purpose of the presentation.

Effective chart *design* will help audience see and understand. Charts should not be crowded. Complex concepts, such as flow charts, should be broken down into several charts, transitioning to an ever more complete picture of the process. Fonts should be large and readable. If a lot of material has to be presented in a single chart, it is useful to have a handout for the audience. If the audience cannot read from the screen and has no handout, they will not be able to follow the ideas.

Presentation software has greatly simplified the preparation of sophisticated and effective presentations. Features such as standardized slide formats, background templates, and entire presentation templates are easily used. The disadvantage of these design tools is that they tend to channel the speakers thoughts and ideas into a predetermined direction. The presenter should make sure that his or her ideas are what are reflected in the presentation, not the presentation software format.

COMMUNICATIONS – MAKING EVERYONE UNDERSTAND
By Michal Safar

Interactive Communications with Some Structure – Group Meetings

Interactive communications are less structured, but not unstructured. They anticipate dialogue and the give-and-take of ideas. This section will provide guidelines for planning and running a successful group meeting. The difference between a presentation and a meeting is that a meeting does not have a speaker and audience. It has participants, and the intellectual content is expected to come from all of the participants. However, individuals do not spontaneously decide to have a meeting. There is usually a meeting leader. In meetings designed to communicate technical information, that individual is usually the technical manager.

Before the decision is made to hold a meeting the meeting organizer should take a hard look at the reasons for having a meeting. Meetings are expensive. They eat up staff time and frequently involve travel and travel related expenses. They create lost opportunities for productive work at the expense of empty discussion. The meeting organizer should try to decide if a telephone call or e-mail would accomplish the objective. The primary reason for having a meeting is that a topic or issue needs discussion, input, and feedback from several individuals. If that is the case, then the meeting is the most efficient way to proceed.

> *A meeting involves discussion, input, and feedback from several individuals.*

There are a number of wrong reasons for holding a meeting. One of them is giving the appearance of being productive. If there is a group of people sitting together in a room the assumption is that they are working. This is not always the case. Another wrong reason for holding a meeting is to appear important. If the meeting organizer can take up the time of a whole group of people, he or she must be important. A third wrong reason for having a meeting is to avoid working. Individuals who spend all of their time in meetings don't have any time to do other productive work. The manager should avoid these pitfalls and only hold meetings when necessary.

There are various types of meetings. *Planning* meetings focus on gathering information from the participants, organizing the information and developing objectives. *Reporting* meetings present information to the group for discussion, problem solving, and decision making. *Administrative* meetings, such as staff meetings, deal with departmental and not necessarily technical information. The information in this section applies to planning and reporting meetings where the intent is the exchange and discussion of technical information.

> *...determine the purpose, participants, time and length...*

Meeting Preparation

Good preparation can insure that a meeting is productive and efficient. There are a number of things to be done in advance. The first is to determine the purpose of the meeting and decide who should be asked to participate. The time and the length of the meeting should be decided and arrangements for a meeting room should be made well in advance. Participants need to be contacted and their attendance confirmed. The agenda should be prepared and distributed in advance along with any pre-meeting assignments.

If the agenda is well organized, the chances are that the meeting will be productive. The meeting objective should be clearly stated at the outset and adhered to throughout the meeting, without the introduction of extraneous topics. The organization of topics for discussion should facilitate productive thinking. The best approach is to start with a short, easy topic to get everyone focused. Then the group can transition to the major topic and focus on that for as long as necessary. As time permits, the group can conclude with subsidiary topics. The agenda should allot a specific time to each topic with the understanding that timings need to be somewhat flexible. The participants should all receive the agenda well in advance with clear instructions on what is expected in the way of preparation by the participants. If some individuals have the responsibility for specific agenda items, they should be made aware of the assignment in advance and agree to it.

COMMUNICATIONS – MAKING EVERYONE UNDERSTAND
By Michal Safar

Meeting logistics can be critical to the success or failure of a meeting. The meeting location should be easy to get to. If individuals are traveling from outside, they should have clear directions for getting there. The meeting start time should be set to accommodate the participants. Nothing derails a meeting faster than having participants come late. The meeting room should be prepared in advance and be arranged such that all the participants can see and interact. Arrangements for water stations, projection equipment, flip charts, and visual aids should be ready in advance with all equipment tested and in working order. The agenda needs to include time for breaks and lunch, and those times need to be adhered to. If there is no schedule, people will leave whenever they need more coffee, a restroom break, or a cigarette. It is much more organized to have everyone break at the same time for a fixed time than to have people wandering in and out. Finally, if the meeting runs over the lunch period, it is more efficient to arrange to have the food brought in and conduct a working lunch rather than taking a break for an outside lunch. This keeps everyone focused.

> *...prepare supporting technical handouts in advance.*

Meetings with a technical focus usually require some kind of handouts or material to be reviewed during the meeting. At a program review this might be the latest progress report. At a planning meeting it might be specifications or other data. The meeting organizer needs to determine what supporting technical material is required and insure that it is available at the start of the meeting.

Conducting the Meeting

It is the responsibility of the meeting organizer to conduct or facilitate the meeting. He or she is the person who identified the need for a meeting and it is up to them to make the meeting a success. Just as with the other communications we have discussed in this chapter, the meeting has a beginning, middle (technical discussion) and a conclusion. Following are some guidelines for conducting a successful meeting.

MANAGING TECHNOLOGY
DEPENDENT OPERATIONS

> *The meeting coordinator should review the agenda step-by-step.*

Start the meeting promptly. It is the responsibility of the meeting coordinator to start the meeting on time and keep it on time. The participants should be introduced individually and the purpose for their presence identified. It is important to involve everyone present in the process. Next the coordinator should review the agenda step-by-step and make sure everyone understands what they are expected to contribute and accomplish. Finally, he or she needs to establish any ground rules. This includes the time for breaks and lunch and other meeting requirements, i.e., no cell phones, no smoking, etc. If the meeting gets off to a good start with all of the participants present ready to contribute, in agreement about the meeting objectives, and understanding the rules, the meeting should be productive.

> *... summarize the discussion...*

Keep the discussion focused and on time. The meeting coordinator is responsible for directing the technical discussion, based on the topics defined in the agenda. He or she should be focused on getting the most out of the participants, making sure that the discussion progresses to decisions and actions, and insuring that the decisions and actions are documented. The coordinator should gracefully discourage conversation not relevant to the topic. He or she should elicit participation from everyone and encourage different opinions. The coordinator should be ready to mediate differences between participants before they become disagreements. To keep everyone focused, the coordinator should regularly summarize the discussion and identify decisions and actions, specifically at the conclusion of each topic. Documentation of the meeting discussion and conclusion is important. The meeting organizer should utilize tools such as flip charts and white boards to capture the discussion. The organizer should have an individual designated to take formal meeting minutes throughout.

End the meeting well. The coordinator should summarize each agenda item with the discussion and decisions. He or she should identify the resulting task assignments and assign those tasks, with due dates, as appropriate. Finally the he or she will coordinate with the group to set the time and place for the next meeting, if needed.

Documentation and Follow Up

Documentation of the meeting provides a roadmap for the participants and allows non-participants to understand what was discussed and decided. All meeting minutes should include the following:

- Date, time, location
- Subject
- Participants
- Discussion by agenda topic
- Resolutions, decisions
- Action items, individual assignments, and due dates
- Date and time of next meeting

Meeting notes should be distributed promptly via e-mail. If there is an internal web site, the minutes can be posted there for wider distribution. Follow-up on individual action items is the responsibility of the manager who coordinated the meeting.

Participating in Meetings

Most managers probably have to attend more meetings than they want. There is seldom a choice; most meetings are a requirement because they directly affect the manager's department. Insofar as a manager can avoid nonessential meetings he or she should do so. However, when required to attend, each participant does have the responsibility of insuring that they contribute constructively. Some guidelines for constructive participation in meetings are:

MANAGING TECHNOLOGY
DEPENDENT OPERATIONS

- *Pay attention.* This is the first and most important constructive contribution a participant can make. Nothing wastes more time than having to repeat the discussion for someone who wasn't listening.
- *Participate.* Everyone attending a meeting should try to contribute new ideas and raise relevant questions.
- *Cooperate.* A meeting is by definition a group discussion. Participants need to build on suggestions made by others and yield the floor when appropriate.
- *Support.* Consensus-building helps to accomplish the meeting objectives more quickly. This can be accomplished by agreeing with good contributions made by other participants and avoiding unnecessary disagreements and criticisms.
- *Stick to the topic.* Nothing is more distracting that an individual who talks excessively, who constantly changes the subject, and who engages in private conversations on the side. Participants should cooperate to keep the discussion focused.

A successful meeting is a cooperative effort among all participants, both those who organize the meetings and those who participate in them.

Understanding Technical Information

Up until now the focus has been on giving managers in technical areas the knowledge and skills to effectively communicate to others within the organization. However, communication is a two-way street. Not only must the manager have the skills to make others understand the objective, he or she must have the skills to understand what others are trying to communicate.

> *Technical information tends to be exclusive...*

Technical information tends to be exclusive, that is, only a discrete group of individuals are experts in a given technical area. Conversely, no individual is an expert in every technical area. Managers in technical organizations are always in the position of having to understand a wide variety of technical areas, some or all of which may not be within their areas of expertise. Following are two approaches for acquiring an understanding of unfamiliar technical information.

Listen and Ask Questions. The quickest way to get smart on a subject is to ask an expert. The technical manager needs to identify the individual within the organization that has the information and ask for help. A good listener can learn a lot very quickly by establishing a rapport with the expert, asking questions, and paying attention.

Learn. Part of the learning process is the effort it takes to locate and understand the information. There may not be a resident expert for the required subject, which means doing research. The earlier section on finding technical information identified the locations within an organization where technical information is located – libraries, computer networks, etc. Using the research tools discussed earlier can help the manager acquire the needed expertise.

Understanding technical information is all about taking the time to find out. Unfortunately, knowledge does not work by osmosis; it requires an effort to acquire it. A truly successful technical manager will develop the skills for understanding technical information as well as communicating it.

Summary

This chapter has tried to present an overview of the means and methods for communicating technical information within the organization. The overview has been directed at managers of technical departments and focused on the types of communications internal to the technical organization. Specifically it covered:

- What technical information is and where it is found
- How to organize technical information
- How to compose technical documents

MANAGING TECHNOLOGY
DEPENDENT OPERATIONS

- Structured communication of technical information
- Structured interactive communication of technical information
- Interactive structured communication of technical information
- Understanding technical information

This chapter needs to be viewed as an introduction to the material presented. More in-depth coverage of these topics can be found in the further reading list.

Finally, one last thought on how to communicate successfully from Alice's Adventures in Wonderland:

'Begin at the beginning,' the King said gravely,
'and go on till you come to the end: then stop.'

For Further Reading

E Writing: 21st Century Tools for Effective Communications. Booher, Dianna.

The Elements of Business Writing: A Guide to Writing Clear, Concise Letters, Memos, Reports, Proposals, and Other Business Documents. Blake, Gary; Bly, Robert W.

The Elements of Style, Fourth Edition. Strunk Jr., William; White, E.B.; Angell, Roger.

Find It Online: The Complete Guide to Online Research, First Edition. Schlein, Alan M.; Kisaichi, Shirley Kwan.

Handbook of Technical Writing. Brusaw, Charles T.; Alred, Gerald J.; Oliu, Walter E.

How to Write Reports and Proposals. Chan, Janis F.; Lutovich, Diane.

Loud and Clear: How to Prepare and Deliver Effective Business and Technical Presentations. Morrisey, George L.; Warman, Wendy B.

The Skilled Facilitator. Schwarz, Roger.

Technical Communications: A Reader-Centered Approach. Anderson, Paul V.

Chapter 7, Lead-in:

RECRUITING THE RIGHT PEOPLE

The challenges brought about by competition based on technological advantage and the skill of the workforce makes recruiting and training critical. Operations managers have to set high standards. Outsourcing may provide temporary options, but these will be short lived if only based on expediency. It is not a cliché that people are truly the greatest asset. This chapter takes a pragmatic view of the operations manager. It claims that highly qualified people are the greatest asset and give tips and tools of how to identify them. Its implied proposition, to hire slowly and to fire quickly is not unique. Many successful operations have adopted this practice. Some say it is the most just and realistic approach. The alternative is costly and may be damaging both to the employer and employee.

It needs to be said that where circumstances permit, recognizing individual differences, training and re-training should be considered. A chance for a behavior change has to be given - note the motivational tools of Chapter 4. After all, there are a number of organization developments with highly successful outcomes. The teachings of Edward Deming, for one, include training of the staff for a change in attitudes. His total quality approach in the long run has turned a number of organizations around. It simply takes time.

CHAPTER 7

RECRUITING THE RIGHT PEOPLE

Tim Ryan

It's no secret that project management, especially technical project management, can be a very delicate, difficult task. Getting your entire team to work together to meet deadlines, produce quality deliverables, and satisfy both your expectations and those of a client can be an extremely challenging process. In fact, how often does dealing with the people on your project team become the most challenging part of your job? Has one of the following situations ever happened to you?

> *...does dealing with the people on your project team become the most challenging part of your job?*

One of your technical people working on the project, though extremely talented in his/her field just cannot work effectively with the rest of the team. How much time did you spend trying to develop a better sense of teamwork and community in this individual?

Has the client ever complained that they "just can't understand" the technical jargon one of the members on your team always uses? Did you try to develop his/her communication skills so that he or she could more effectively communicate with the client? Did it work?

RECRUITING THE RIGHT PEOPLE
By Tim Ryan

Some "natural" managers have the skills to handle the people on their team and the demands of their bosses as well as the relationship between their organization and that of the client's. Yet some of these people just don't have the "techie" background to skillfully comprehend all the terminology involved in such relationships? Are you one of these managers?

If you answered "yes" to any (or all!) of these situations then you'll find some of the ideas in this chapter extremely useful. Many times the problems executives experience in communicating with their team members can be addressed before they ever occur.

By focusing on selecting the right people for your team before a project even begins, you cut off many future problems at the pass. Before we get in to this, though, let's build a strong foundation for these ideas and then skillfully use that foundation as a starting block for building a successful project management team.

Behavior is the Key

As an executive or manager of projects for your company, chances are you've spent a few years in your profession and have worked hard to get here. Think back over to some of your most successful projects. Now think of the people involved. What were they like? How would you describe them? Now think of some of the more trying experiences you have had as a project manager. What was the team like in these instances? How do they compare to the group of individuals you worked with on your most successful endeavors?

Perhaps you can already see that some of your best experiences in project management came when you were working with great people. Of course, this isn't very surprising information. When you work with great people, you get great results. So many executives agree that people are their most important asset, yet so few focus on making sure they have the right players on their team before beginning a project. If so much time is spent in planning the project so that the best results can be achieved, isn't it

MANAGING TECHNOLOGY DEPENDENT OPERATIONS

just as logical to make sure that the team you're working with can actually go out and execute these plans?

> *The first step in making sure you've got the right people...*

The first step in making sure you've got the right people on your team involves the basic tenets of Competency Theory. Plainly put, an executive should study his/her best people in a certain position. Almost every organization has some true "Difference Makers" on their staff, the ones your organization would love to have more of. Who are these individuals? What makes them great?

In most cases, it's their behavior that makes them great. We can define Behavior as *what motivates a person, how they think, how they act and how they interact with others.* The bottom line is that when you work with an individual, you are basically renting or leasing their behaviors. Therefore, it' very important to make sure that the individual you want either to fill a position on your team, or the person currently holding that position on your team, has the right behaviors to get the job done.

> *By the time we reach the workforce, most behavior has already been permanently molded...*

Now the next critical idea to consider is pivotal to understanding the upcoming ideas: Behavior modification is ineffectual, short lived and in many cases, impossible. Think about it. Have you ever told one of your employees that he or she has got to be a better team player? In the long run were they able to effectively change into a person who works well with others and puts the goals of the team above his or her own? Or did they make an effort for awhile and then fall back into old habits? Too often the areas we try to develop in an individual are really just part of their behavior. Some may think that behavior modification is possible, many times offering the example of a long time smoker who finally quits. However it was not their behavior that changed, just a habit. Behavior is the lowest common denominator. We all exhibit certain behaviors that make up who we are. By the time we reach the workforce, most of that behavior has already been permanently molded through our home life and schooling, and is, for the most part, set in stone. Even if you could effectively change a person's behavior, the amount of time and energy that would have to be spent to truly accomplish such a feat is time and energy that is being taken away from actually managing the task at hand and therefore watering down the level of quality. That, however, is the negative side of the fact that behavior does not change. The positive side is that if an individual has exhibited positive behaviors in the past, they are very likely to continue exhibiting these behaviors in the future. Why? Because behavior doesn't change!

It's high time to get out of the behavior modification business and into the behavior selection business. Selecting your team based on the desired behaviors necessary for a position also opens up your search. If someone has the right behaviors, and the necessary requirements to get the job done (i.e. programming experience, Ph. D. in chemistry, etc.) we no longer focus on the superfluous such as previous work history. It is also a very fair way to select people since we are focusing primarily on behavior, regardless of race, orientation, sex, or creed.

Now that we understand the importance of behavior and its predictive value on future success, we can combine this idea with the aforementioned tenets of competency theory to understand the concept of building Behavioral Benchmarks. A Behavioral Benchmark is a standard against which people in your organization may be measured. The Behavioral Benchmark is a valuable tool to help you study the best in your organization and industry, and determines what behaviors a person in a given position must have to be successful. It establishes a standard and objective measurement that minimizes emotional decisions. Once the Benchmark is in place, employee or candidate profiles are measured against this detailed behavioral picture of the position and the odds of better decisions are significantly increased.

MANAGING TECHNOLOGY
DEPENDENT OPERATIONS

Of course, it takes time to create Behavioral Benchmarks but the investment of time up front to make sure a potential or current team member has the "right stuff" is well worth it. At some point during the project you *will* invest that time. The question is whether you want to invest it researching the position before the project gets underway, or spend it coaching, counseling and troubleshooting once the project is underway. Consider Figure 1:

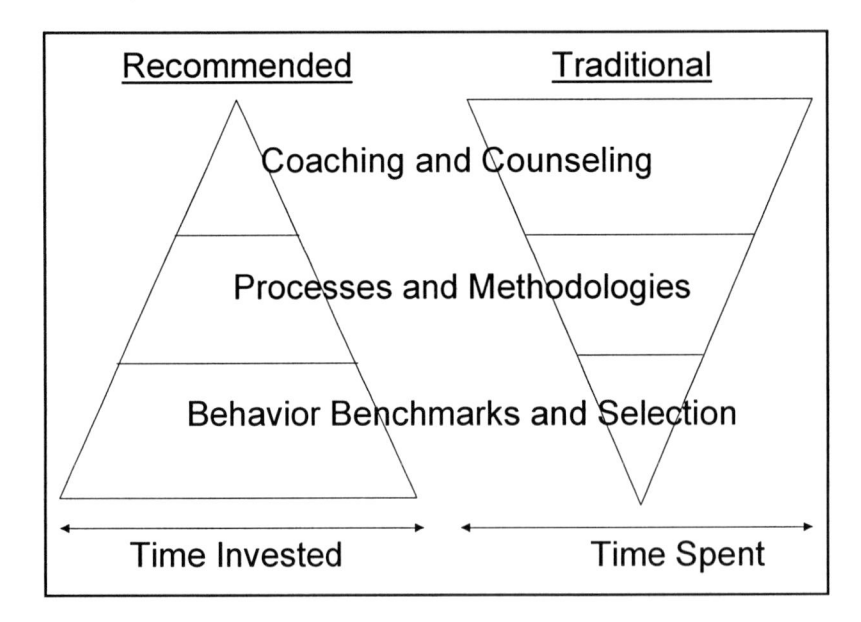

Figure1: Selection Management Systems

Recommended	Traditional
Coaching and Counseling	
Processes and Methodologies	
Behavior Benchmarks and Selection	
Time Invested	Time Spent

This Chart demonstrates the difference between traditional staffing methodologies and the approach that we recommend. Traditionally, people are hired almost solely on what their resumes may say. Yet, some 30% of resumes have lies in them, and even the completely honest ones do not say enough about their behaviors. The normal interviewer may have some great questions but candidates read books, attend seminars, or take classes on how to interview well. Chances are they've heard this type of question before and have already learned to give an answer similar to the one you were told

RECRUITING THE RIGHT PEOPLE
By Tim Ryan

to look for. Similarly, these questions many times just test what a person knows, and as Brian Tracy says, "You get hired for what you know but fired for who you are." Who you are is your Behavior! This means that managers have to take the time to think of and construct the behavioral benchmarks necessary to improve their chances of selecting the right individual, promoting the player with the most potential and retaining your "Difference Makers" and "Everyday Heroes". This chart demonstrates a simple but very powerful truth. Managers who do not take the time to construct behavioral benchmarks for the positions on their team and then make staffing decisions relative to that standard will inevitably spend that time coaching and counseling an individual to try to get him or her to perform in the desired fashion. When an employee who works on a project management team isn't working out, the first question good managers usually think of is, "How can I help this individual perform?" That's the wrong first question, though. The first thing a project manager should ask when a team member is not working out is, "Does this individual have the right behaviors to get the job done?" If the answer is "no" to the second question, then all the coaching and counseling in the world will not develop this person into something they are not.

> ... construct the Behavioral Benchmarks necessary to improve the chances of selecting the right

There is a fine line here, however. It is important to think in terms of behaviors, the characteristics that have been "hardwired" into who we are. It's okay to coach and counsel an individual with the right behaviors in order to develop their skills in a certain endeavor. For example, it makes sense to work with an employee who is capable of learning new skills such as different software programs or operating methodologies. It does not make sense to send an individual who is not assertive to seminars that guarantee to teach assertiveness and intensity. The individual may come back invigorated and for a few weeks may actually seem more assertive, but as time passes their true colors come shining through and the same person you knew before the training seminar is once again back and as mild as ever. Consider Figure 2.

MANAGING TECHNOLOGY
DEPENDENT OPERATIONS

Figure 2: MPR Competency Model

Rank Ordering of Combinations

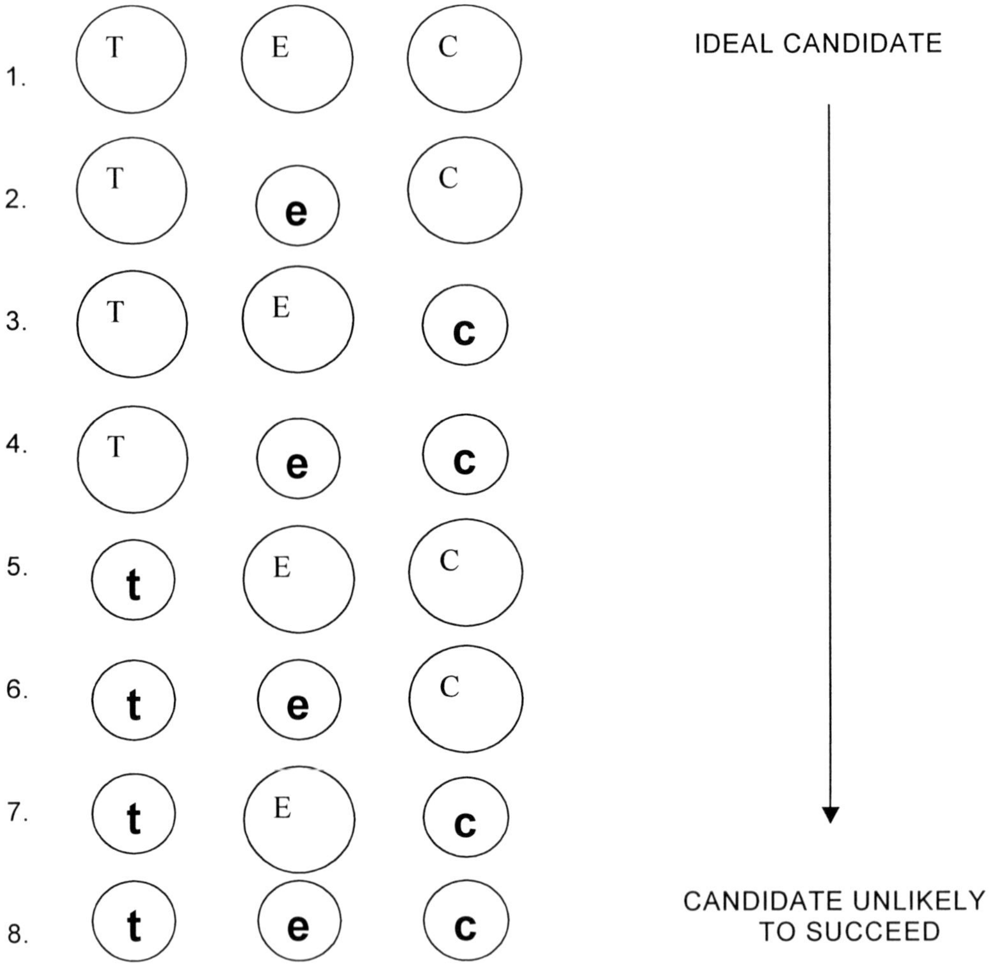

1. (T) (E) (C)   IDEAL CANDIDATE
2. (T) (e) (C)
3. (T) (E) (c)
4. (T) (e) (c)
5. (t) (E) (C)
6. (t) (e) (C)
7. (t) (E) (c)
8. (t) (e) (c)   CANDIDATE UNLIKELY TO SUCCEED

T	=	Talent — Behavior Traits required to get the job done.
E	=	Experience — Job related experience, education, and training that contribute to greater productivity.
C	=	Chemistry — Personality that fits into company culture, the manager, and the work group.

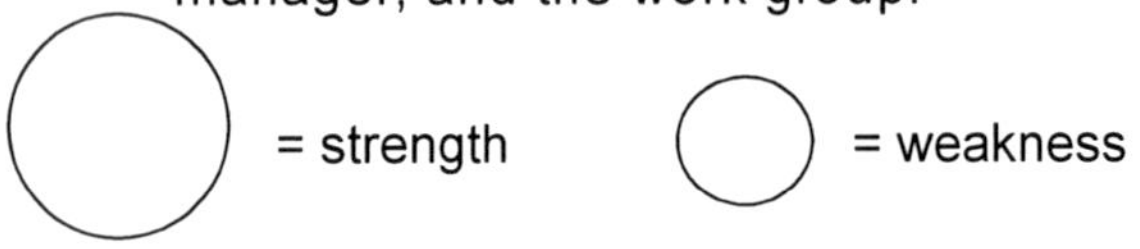

(large circle) = strength   (small circle) = weakness

RECRUITING THE RIGHT PEOPLE
By Tim Ryan

Consider the preceding diagram. In this diagram the "T" stands for the traits necessary to get a job done. We also refer to this "T" as talent because a collection of desirable traits necessary to get the job done can be called talent. The "E" stands for a person's education and experience. The "C " stands for chemistry. Chemistry is a group of traits that are a part of your company culture.

> *If someone has the talent and the chemistry to fit the culture of your team, they can learn the necessary skills*

In building your project management team, you should consider this chart. For any team member currently on your team, or someone that you would like to add to the team, no one should fall below the #2 distinction. If you are going to give on anything, give on education and experience. If someone has the talent to perform a job well and the chemistry to fit the culture of your team and company, they can learn the necessary skills to enhance their effectiveness.

Let's take a look at some of the other numbers and discuss some potential reasons why they may prove to be team members who are "less than the best":

- Ranking # 3, Strong Talent, Strong Education/Experience and Weak Chemistry: This individual may know everything necessary to perform a particular job and even have the traits/talent to perform it extremely well, however if he or she cannot work well with his or her teammates because of a lack of chemistry, you will inevitably run into problems as the manager. Instead of managing the project you may spend more time managing interpersonal issues of the team. In fact, most turnover is the result of a lack of chemistry. Remember, "You get hired for what you know and then fired for who you are."

- Ranking #5, Weak Talent, Strong Education/Experience and Strong Chemistry: Individuals who fit this description can be very dangerous to managers. Although, they simply do not have the talent to ever perform as well as you need them to, a person like this has all the experience, education and chemistry to fit right into the organization and probably be well liked by his or her peers. At some point, their lack of talent will force you to remove him or her. Yet because of their strong chemistry with their team members and co-workers, as well as their education experience, the manager that makes this decision may earn a negative reputation that may hinder or interrupt successful management and performance.

- Ranking #7, Weak Talent, Strong Education/Experience, Weak Chemistry: Although it seems obvious not to select such an individual for your project team, there are many people like this who still work and disrupt project management teams. Why? Many times it's the result of an excellent resume. An important idea to keep in mind is that resumes are balance sheets without liabilities. This individual's resume may look impressive and they may even interview well, but once they become a part of your team, it's usually apparent that he or she cannot perform. As a manager you'll be forced to deal with this person and thereby risk losing the productivity of the rest of your team not to mention their respect for you as a manager.

Hopefully, this shows why "settling" on candidates can have disastrous results in the long run. To successfully take on projects and complete them, a manager needs a great team of people who have both the talent and the education/experience to get the job done right and the chemistry to work with their teammates toward a common goal.

Building the Project Team

Before you take on a new project ask yourself, "What are the different positions I need to fill to put together a great project management team for the upcoming job?" Once you've put together a list of the different positions necessary for a well-rounded team, you should then think of the best people who currently hold this position. Sometimes, though, there may not be a

RECRUITING THE RIGHT PEOPLE
By Tim Ryan

real "Difference Maker" that currently fills the position in the company, and that's okay. There's a good chance that in the experiences gained from other projects you handled, either with your current or past employer, you've run into a few good men and women who held the position and performed well in it. Think about them. These are the individuals on which you will be basing your Behavioral Benchmark.

On a blank sheet of paper, write the name of the position and then the name of the individual that is the best in this job. Then, list out all the different ways that you can think of to describe this individual. You can use examples from their professional or personal life to describe them (Remember, behavior is consistent and does not change regardless of whether it's at work, at home, in the community and so forth).

> *...write the name of the position... then particular traits...then the individual that is the best in this job...*

After you've compiled this list refer to the Glossary of Behavioral Traits (Appendix). Review the list of adjectives or phrases you used to describe the individual. Then compare these descriptions to the definitions of the different behavioral traits in the glossary of traits. You'll notice that the adjectives that you used to describe the individual will first fall into one of the four categories that behavior can be broken down into:

- Motivations,
- Modes of Thinking,
- Modes of Acting
- Modes of Interacting.

After you've broken your descriptive list into these areas, continue to sort these adjectives and phrases until you can put each one under the heading of a particular trait. It's okay to have more than one of these phrases or adjectives under the umbrella of a particular trait. Remember, behavior can be considered a lowest common denominator, so it's not unusual that a few of these descriptors all describe one trait. If anything, it should let you know just how critical that trait is to the successful fulfillment of the position. Example:

>Position: *Programmer*
>Name: Emily Smith
>Description:
>1. Hardworking
>2. Puts in long hours
>3. Very knowledgeable of code
>4. Enjoys working well with others
>5. Committed to delivery
>6. Excellent researcher
>7. Energetic
>8. Proactive
>9. Friendly

After reviewing these descriptions I would then compare these to the glossary in the Appendix . To keep it simple, we first look at which of these descriptions fall under *Motivations*: Number 5 – committed to delivery. This is an example of the Mission of Service Trait. Numbers 1, 2, 4, 6 and 8 can fall under *Modes of Acting*. Once we have figured out which category these traits fall into, we can start designating which trait these words and phrases describe. Numbers 1 and 2 are indicative of the Intensity trait, while number 3 corresponds to Mastery. Number 8 indicates the presence of the Proactivity trait, and number 6 is an example of the Researcher trait. Finally, Numbers 5 and 9 fall under the *Modes of Interacting* category and both are an example of the Relator trait.

Thus far we have discovered the behavioral traits that are critical for the success of the position by describing Emily, a high level performer in that position. By doing this we have created a Behavioral Benchmark of Emily Smith. The next step is to think about the position of Programmer for your company, (do not necessarily think of Emily anymore) and rank these traits in numerical order, with number one as the most important and then consecutively downward from there.

RECRUITING THE RIGHT PEOPLE
By Tim Ryan

When comparing two traits ask yourself, "Which one of these traits is more important for this job, on my team, in my company?" or, "If I have two candidates sitting in front of me, one of whom exhibited more of one these traits and one whom exhibited more of the other, which candidate would I pick? Why?" Keep in mind, however, that by picking one over another, you are not *losing* either trait from your benchmark. You are simply recognizing which traits are going to be called into action more often. The goal of doing this is to challenge your pre-conceived notions of this position and it's generic job description and requirements. We've used Emily to see what traits are so important for successful execution of the job *in general*. Now we want to *think of the position itself*. Which traits are going to be absolutely critical *for this position, on your team, in your company*.

> *Your top traits can be considered "deal breakers"*

By ranking these traits in numerical order you will end up with an accurate picture not only of what traits are necessary for the successful fulfillment of the position but also in what proportion. Your top traits can be considered "deal breakers." If a candidate does not exhibit these traits, they most likely won't work out well in this position because it is precisely these deal breaker traits that are going to be called upon regularly in order for this person to perform well.

Surfacing The Behavioral Traits

Now that you've built your Benchmark, it's time to start using it to make informed staffing decisions. If you are currently interviewing new applicants for a Benchmarked position, you'll need to construct an interview.

First, take out your Benchmark and using the glossary of traits, review the definitions of each trait included in your benchmark. Once you feel comfortable with the definitions, create open-ended, situational questions designed to surface that particular trait in an individual. Ask questions that look for positive, specific examples of the candidate exhibiting that trait. Such

MANAGING TECHNOLOGY
DEPENDENT OPERATIONS

> *...create open-ended, situational questions...*

questions usually begin with, "Tell me about a time when," or "Can you give me an example of".

Example: Motivations, Mission of Service Trait : *Tell me about a time when you went the extra mile for a customer.*

Sample Answer 1: "Well, you have to go the extra mile for the customer every time. If you don't then someone else will and the minute that happens, you could lose the customer. Given how long it takes to find one customer, you simply can't afford not to go the extra mile, so I always do."

Sample Answer 2: "One time one of my smaller accounts said they we're going to call me at 4:30. I waited around until 5:30 when they finally called. It turned out they needed a new part for one of their systems. All the delivery drivers had already gone, so I signed the part out myself and drove it to the customer's sight. I didn't get home until 8 o'clock that night, but by the time I left the company I was working for at that time, that account had grown into one of my biggest."

Which of these answers shows, clear, positive evidence of the Mission of Service trait in the candidate. If you said Sample Answer 2, you're right!. Someone who can easily recall actual examples of a behavior shows that particular trait to be a part of who they are. If they've exhibited this trait in the past, then they are very likely to continue to exhibit in the future. (Behavior is consistent and unchanging!)

This Behavioral Benchmark is not limited to hiring only, though. This Behavioral Benchmark can be used for any staffing decision made, regardless of whether it's to hire, fire, promote or retain. You can use this as an objective standard, an instrument in your executive tool box. It will not only significantly reduce the amount of purely emotional staffing decisions you make, but also increase your chances of staffing a position with the best possible person.

RECRUITING THE RIGHT PEOPLE
By Tim Ryan

In order to use the Behavioral Benchmark for such staffing decisions as promotion, you can try one of two approaches. Before promoting a candidate internally into a position, first check the Behavioral Benchmark. Review the traits necessary for the successful fulfillment of the position. Now think of the individual whom you wish to promote. Compare their behaviors to the ones listed in the Benchmark. Do they measure up?

The second approach is to simply put the individual through the interview you've created to correspond with the Behavioral Benchmark. If the candidate can provide positive examples of the requisite behavior, they may be the right person for the job. However, do not be too lenient with an individual. You will have to live with your decision for some time. It's pretty rare that "settling" on a candidate produces the desired results.

> *You will have to live with your decision for some time... through the interview you've created!*

In the same fashion, you can use the Behavioral Benchmark for such staffing decisions as retention or termination. When determining a current team member's future on your team or perhaps with your organization as a whole, you can use the Benchmark as a guide. Simply observe their behavior and ask others for examples of their behavior and compare them to the Benchmark. If you've spent the time to create this accurate objective standard for measuring a person who will successfully fit with your company, and a person on your team does not measure up to that standard, you will have to take appropriate action for the sake of the project, your team and your responsibilities as a manager. However, this does not necessarily mean that the individual has to be terminated. If you've observed enough of their behavior, you can compare them to some other benchmarks for positions they may be good at. For instance, someone who works hard but simply isn't working out in sales may have all the behaviors to be a great customer service representative.

Conclusion

The people that make up your project management team will determine not only whether the project is successful but also whether *you* are successful as a manager. Investing the time to create Behavioral Benchmarks as objective standards will be extremely useful for putting together new project management teams as well as assessing your current team. The bottom line is that even with the best equipment, the best, strategy, the best planning and the best clients, without the right people you're destined for difficulty. As a manager, so much thought goes into strategic planning for products services, systems and projects themselves. Shouldn't just as much thought be put into the people you expect to go out and execute these tasks? Build your future with the best!

Figure A-1

Appendix: GLOSSARY OF TRAITS

MOTIVATIONS	Behavioral Traits that address the fundamental "drives" of an individual and are characterized by more than the simple desire to earn money to satisfy basic necessities; that is, what provides the individual with fulfillment through work activities. (At least one of these 6 Traits must be selected as critical)
ACHIEVER (Individual)	• Confident and self-assured • Seeks independence and recognition • Driven to high levels of accomplishment
COMPETITOR (Individual)	• Energized by competition • Responds to measurable performed goals • Driven to produce
MISSION OF SERVICE (Individual & Management)	• Service oriented • Team player • Committed to family and community
PRODUCER (Management)	• Results oriented • Seeks objectives • Measures performance throughout the unit
RESPONSIBILITY (Individual)	• Conscientious and dependable • Good attendance and punctuality • Committed to delivery of tasks
TECHNICAL MASTERY (Individual) (for technical positions only)	• Committed to continual self-education • Large store of industry knowledge • Intrigued by new developments
MODES OF THINKING	Behaviors that address an individual's capacity to gather information and process it (sort, parse, problem-solve & analyze). Additionally, it looks at an individual's ethical principles, as well as, their creativity, flexibility, and adaptability.
DECISION MAKER (Management)	• Thorough research • Aware of parts-to-whole relationship • Strives to harmonize competing interests
DISCERNER (Individual)	• Skilled in self-appraisal • Quickly sorts the critical from the superfluous • Acts appropriately
INNOVATOR (Individual & Management)	• Constant search for better methods • Flexible and adaptive • Encourages new ideas
VALUES (Individual)	• Integrity and honesty are hallmarks • Refuses to cut corners or over-promise • Represents company judiciously
MODES OF ACTING	Behavioral Traits that address an individual's approach and skills for accomplishing work functions. These would include organizational and time management skills, planning and prioritization, as well as, initiative, work focus, and physical and mental stamina. These are essentially job-related, functional Traits.
ARRANGER (Management)	• Deploys resources effectively • Concerned with efficiency and streamlining • Long-range planner

MANAGING TECHNOLOGY DEPENDENT OPERATIONS

CULTIVATOR (Management)	• Forward-looking • Aware of company position in the marketplace • Quality conscious
DEVELOPER (Management)	• Astute selection decisions • Personal attention to training and coaching • Sense of satisfaction in growth of subordinates
INTENSITY (Individual)	• High stamina and endurance • Focus on work activities • Active hobbies
PROACTIVITY (Individual)	• Looks for solutions • Initiates change and improvement • Doesn't blame or shirk responsibility
PROSPECTOR (Individual Sales)	• Targets accounts carefully • Continually probe and penetrate accounts • Methodical assessment of "fit" between prospect and the company
RESEARCHER (Individual Technical & Systems)	• Full range of information gathering techniques • Methodical hypothesis formation and testing • Skilled troubleshooter
STRATEGIST (Individual)	• Well organized and skilled in prioritizing • Well developed grand planning skills • Brings tasks to closure
TECHNICAL MASTERY (Individual) (for non-technical positions only)	• Committed to continual self-education • Large store of industry knowledge • Intrigued by new developments
MODES OF INTERACTING	Behaviors that address an individual's interpersonal skills, that is, how they influence, interact, and get along with others.
AFFILIATIONS (Individual Technical & Systems)	• Enjoys sharing expertise with other professionals • Contributes to team bonding • Taps into knowledge in a variety of fields
ASSERTOR (Individual)	• Straightforward and direct • Opens doors, closes deals • Sense of drive and aggressiveness
COMMUNICATOR (Individual)	• Adapts to level and interest of others • Confident public speaker • Articulate conversationalist
EMPATHY (Individual)	• Sensitive listener • Recognizes role of human nature in business decisions • Need to reach out and comfort
MOTIVATOR (Management)	• Stimulates enthusiasm • Creates "buy in" from employees • Utilizes incentives and praise generously
PERSUADER (Individual)	• Skilled listener who identifies motivations • Probes and questions others' agendas • Strives to influence others
RELATOR (Individual & Management)	• Outgoing and congenial • Takes time to engage associates on a personal level • Promotes harmony and positive relationships

Chapter 8, Lead-in:

MANAGING DIVERSITY – TECHNICAL AND CULTURAL

One key premise of this book is that there is diversity between functions and people's background that needs to be addressed. General management must understand the cultural and professional biases that exist within their organization. The more technology dependent the operation, the more the bias and differences are noticeable. The spread between the people, processes, and things gets bigger. Functional "silos" and comprehension of ethics where technology decisions are involved get more complicated.

Beyond their functional roles, there obviously is the diversity in the people themselves by their gender, seniority on the job, ethnic background, departmental traditions, specialized training, professional discipline bias, or previous employment practices they bring to the job. While all cannot be covered in depth, this chapter provides a number of tools to some of these in a practical sense.

As a complement to the suggestions of Chapter 7 to recruit the right people, this chapter suggests examining individual behavior for its rational and emotional elements as well as the relationship to personality, perception and attitudes. It leads to possible reassignment or training opportunities. Simple instruments are suggested to assess personality-job-fit, job-content-attitude, as well as change and operations knowledge management tips.

Chapter 8

MANAGING DIVERSITY – TECHNICAL AND CULTURAL

Donatas Tijunelis

This chapter briefly addresses diversity issues as they pertain to technology management. It provides food for thought regarding behavior patterns, professional and age bias, diverse concepts of ethics, and factors to be considered in international operations environment.

> *Know when an assignment fits...*

Organizations can be composed of machine operators, in-plant technical service personnel, schedulers and bookkeepers, personnel managers, design engineers, chemists, purchasing staff, loading dock supervisors, logistics personnel, plant supervisors, process engineers, cost accountants, research and development personnel, safety and environmental experts, secretaries, quality control technicians, field service groups, and various project managers and supervisors along the way. Managing operations can mean managing any part or all of the above. It is the management of people, processes, and things. In smaller operations, the same person covers several functions, wearing several hats and taking on several roles. Each role develops its own personality, may require a new attitude and perception skills. In organizations, diversity is inherent. Beyond the technical skills and education, this obviously is in the diversity in the people themselves by their gender, ethnic background, departmental traditions, and other industry practices they bring to the job. Their education/ training, professional discipline bias and the organizational hierarchy in which they see themselves further complicate the situation.

The more technology dependent the operation, the more these differences are noticeable. The spread between the people, processes, and things gets bigger. The specialists are more specialized and less understanding of the generalists. The technician is more technical while the people-manager more socially focused to manage diverse interests. Hiring and retaining specialists is a challenge. The turnover of staff is greater, new hires bring more radical ideas and the diversity in the staff grows. The spread between the senior and junior staff widens. A simple example of the use of computers can easily be appreciated. The young adapt quickly, the older take more convincing and time to adopt new computer tools.

> *The specialists are more specialized and less understanding of the generalists....*

Diversity management in a small organization may not be easy, but its challenge is visible. The line of supervision is shorter. By necessity, people are forced to adopt several roles and adjust their work habits to those of their peers, subordinates, and superiors. Lack of harmony is more obvious. Decisions are more direct. People fit in or get out. As in a family, counseling may be necessary, but it is of the psychological, not so much of organizational nature. It is in the larger companies, with more structured operations, where formal organizational behavior and organizational development tools are needed and find their application.

Individual Behaviors

Individual behavior in organizations has rational and emotional elements. Looking only at some of the very basic components of the nature of individuals may help you to recognize the behavior differences on the job and how to deal with them. On a very basic level, individual differences relate to personality, perception, and attitudes. (See Table 1).

Table 1: Individual Differences on a Basic Level

SOURCE OF BEHAVIOR	CHARACTERISTIC	TIP
• PERSONALITY	Built-in, stable	Know to avoid clash
• PRECEPTION	Specific sensitivity	Seek for alternative view
• ATTITUDE	Opinions, changeable	Check job design fit, train

Personality shows up in the organizational behavior in the unique adjustment people make to their environment. It is generally believed that while personality remains relatively stable, behavior may vary depending on the circumstance (1). Management's challenge then is to recognize the limits. Know when an assignment fits the person and when it is beyond the ability to adopt. A project team composed of people whose personalities clash beyond their ability to adjust is doomed to failure. There are many personality test instruments that organization behavior experts recommend. These have to be used very cautiously since the very fact of taking a test may introduce stress which can be irreparable. For day-to-day management it may be better to simply recognize the personality difference, observe behavior and based on those observations judge how an assignment fits the person.

> *...while personality remains stable, behavior may vary depending on the circumstance...*

There are several factors determining the personalities of individuals. The source of these can be genetic makeup, environment, the immediate situation, and/or social and cultural influences. Genetics govern biological makeup, size, reflexes, etc. Environment, in which a person is raised in terms of security, the need to cope with situations, and independence or interdependence will impact personality. The immediate situation, such as being demoted or promoted, will have an influence. A chance encounter with an unusual career opportunity and its failure or success can be a factor. Finally, the culture in the form of accepted values and norms of the society to which one is accustomed will also effect the exhibited personality.

With respect to technology management, an example may be the reaction people have to a long vs. short-range assignment and the type of projects they favor. People may be passive or active, act maturely or immaturely, be self-confident or timid, be self-aware or oblivious to how they are viewed. As a manager who has an opportunity to assign individuals, consider reviewing Table 2.

Table 2. RANK FOR PERSONALITY FIT WITH TECHNICAL CHALLENGES

PERSONALITY	CHARACTERISTICS & *PREFERENCES*	RANKING High	Med	Low
Realistic	Shy, genuine, persistent, stable, conforming, practical *Prefers physical and skill activities*			
Investigative	Analytical, curious, original, independent *Prefers thinking, organizing and understanding*			
Social	Sociable, friendly, cooperative, understanding *Prefers activities that involve helping / developing others*			
Pragmatic	Confronting, efficient, practical *Prefers rule-regulated, orderly, low challenge activities*			
Enterprising	Self-confident, ambitious, energetic, dominating *Prefers verbally influencing others, attaining power.*			
Artistic	Imaginative, disorderly, idealistic emotional *impractical* *Prefers ambitious, unsystematic activities for self-expression*			

MANAGING TECHNOLOGY
DEPENDENT OPERATIONS

Thoughtful examination of Table 2 should give a fair idea of personality fit for a specific technical project assignment or specific management challenge. Interface with marketing may be a good choice for an enterprising individual. Given the option, do not assign a very exploratory project to one who ranks high as a realist. That person may do better on a project to scale-up a defined product for expanded production. Since you may not always have the choice, at least be aware what to expect from the staff that you have in their respective assignments. Personality profiles established through organizational behavior surveys as in, Otto Kroeger and Janet M. Thuesen's *Type Talk At Work* (2) could be used, given the option, to make the best job assignment for maximum success. Within the context of practical management, the use of such tools is to make the best fit of an assignment and not to seek a personality change.

People also behave differently because they perceive things differently. In very simple terms, not everyone is an artist. Some people are colorblind. People's sensitivities vary. People react differently to different stimuli. An engineer may stay attentive to a long and detailed description of some production process modification while the sales person, placing an order, could not care less. One person sees a process defect while another may see a new feature to advertise. A number of troubling aspects of perception differences arise in work relations, team-building, and project management. Unfair stereotyping, as

> *...perception differences arise in work relations, team building and project management.*

the example just cited, is one. A halo-effect, where an isolated experience is generalized, is another. One's experience on one job is assumed to be applicable in a different case. Another common problem is projecting one's traits onto another or attributing one's own faults to those of others. Finger-pointing, often encountered on the job, can be the result of perception. As a practical matter, there is little that can be done about improving perception in technical or general management except for greater experience and training. With experience, subtleties become apparent. With training and experience, the savvy for business impact grows. Someone who is good at fixing cars is likely to be recognized as technically perceptive in a technical service function. On the other hand, an engineer who has an MBA or ran his own business while in college may be recognized as potentially more "perceptive" for commercial application opportunities of new technology.

While personalities and perceptions may be hard to change, attitudes, and dispositions toward other people or events are learned and can be changed. Attitudes are also opinions. Opinions change. Behaviorists conduct attitude surveys. Surveys to measure job satisfaction are a frequent tool. The most direct approach is an interview.

> *...attitudes are learned and can be changed.*

Attitudes improve as jobs become enriched and more satisfying. It is important to keep in mind that an enlarged job, expanding its tasks and responsibilities, is not the same as an enriched job. An enriched job is one where broader responsibility carries also broader authority – expands autonomy. One factor that influences attitude is the level of enrichment that good job design provides. It may be helpful, therefore, to examine how much job design affects the attitude of a technical staff member according to the generally accepted core dimensions of their jobs (Table 3).

Consider what the technical job offers against the possible inclination of the person. This, in a technology dependent operations environment, could include a data base handler, order entry typist, secretary, engineer, or scientist. A mismatch is likely, in the long run, to affect attitude. Job re-design, re-assignment, or at least some form of appropriate accommodation that satisfies individual needs may have to be initiated.

MANAGING TECHNOLOGY
DEPENDENT OPERATIONS

Table 3: JOB CONTENT AFFECTING ATTITUDE

CORE DIMENSION	PRINCIPAL FEATURES	CONTENT		
		high	med.	low
Task Variety	Different operations to perform *Working on more than one project*			
Task Identity	Performing a complete piece of work *Extent of responsibility and recognition for a project*			
Significance	Work that appears to be important *Technical challenge and / or business impact*			
Autonomy	Control over their own work *Ability to set goals and technical task priorities*			
Feedback	Information about performance *Peer evaluation of technical and / or business achievement*			

To change attitudes in technology-dependent operation's environment, where the gap between technical and non-technical staff is likely to be more apparent, several training approaches can be effective. The training program promoted by Steven Covey (3) has been effective in changing people's attitudes. It teaches self-improvement that works well with professionals. The fact that it does not stop at the recognition of differences, but teaches practical changes in attitude, is of unique value in operations management where inter-disciplinary cooperation is essential. Besides cooperation and a win-win attitude, it leads to striving for synergy and entrepreneurship, which is vital to technical operations with limited resources. Furthermore, it emphasizes the value of honing one's skills to keep up with fast-changing technology. The very basic tool to change attitudes in managing diversity is to instill respect for interdependence and to review the following steps that management can take:

> *...strive for synergy and entrepreneurship...*

1. Understand and Teach to Understand Self-limitations
2. Encourage to Seek Progress
3. Accept and Understand the Capabilities of Others
4. Seek Synergy
5. Appropriately Promote and Accommodate

Dealing With Views of Ethical Behavior

Ethics come into play on the job in many ways. At a simple level, ethics can be viewed as attitudes. These can manifest as respect given to peers, management, or the public at large. With regard to a technical assignment, a purely technical job itself is ideally impersonal. It should have relatively little bias and little ethical issues. Theoretically, technical project teams made up of men, women, and ethnic minorities or any diversity should not be a factor. That said, cultural differences exist with related attitudes toward work assignments. For example, in some international settings, it is considered unethical to assign physical labor tasks to a technical professional. Some foreign engineers and scientists tend to look down on anything that looks like manual work. Cross-cultural training may be needed. There also may be some obvious difference in the case of physical inability and health issues, for which usually there are standardized human resource management policies.

> *...ethics -- from the primitive "rules of the jungle" to the complex needs of a society...*

Technical facts, the laws of physics or engineering principles, are what they are, yet the choice in their application have ethical implications. Attitudes toward public safety come into play. Here the situation gets to be more complex, more subjective. To begin, there are the subtle issues of ethnic-family traditions, religious holidays, or political commitments. Here the values

MANAGING TECHNOLOGY
DEPENDENT OPERATIONS

and traditions of the organization and the ethics of individual managers range from the primitive "rules of the jungle" to a consideration of complex needs of a society. The relationship between general and technical management can be examined according to the following evolutionary stages of ethical management from the primitive to the contemporary:

       Stage 1: Punishment and obedience
       Stage 2: Self-gratification
       Stage 3: Approval of others
       Stage 4: Law and order
       Stage 5: Social contract
       Stage 6: Universal ethical principles

Where do your relationships fit? Ethics between technical people or between general management can be as varied as between technical and general management. Technical people and general managers are equal in their ability lie and cheat or they can be very honest and ethical. Assuming that honesty and ethical behavior is desirable, are general management and the technical staff at the same stage? If not, expect misunderstandings and conflict. Training may be in order.

Universal ethical principles are not always easily understood and accepted. It is worth noting that ethical management can take one of two forms. Each has a different slant, but with the same good intentions. Misreading can lead to finger-pointing, unfair criticism and disharmony. It is therefore good to learn to appreciate the distinction between the two:

- **FORMALISM**....morality on the basis of right or wrong... laws
- **UTILITARIANISM**....for greatest good of the greatest number...heart

The one approach sticks to the rules and insures that bias, favoritism, or other improprieties do not enter, while the other is willing to bend the rules for the greater good under the circumstances. The one can seem heartless, while the other may allow the rules of ethical behavior to drift into inappropriate application under changed circumstances. Both seek to be fair. Consider the examples of some typifying decision alternatives that arise in the following categories:

Table 4: Example of Possible Ethical Decision Alternatives

FORMAL	UTILITARIAN
Adherence to safety rules	Cost-benefit analysis
Establishing hours of operation	Flex-time and special leave options
Advancement of science objectives	Application of technology issues
Environmental protection laws	Economic development opportunities
Patent law enforcement in medicine	Dealing with epidemics and poor
Anti-abortion laws	Woman's rights

The publication of an organization's beliefs, mission and values helps. One good piece of advice may be to consider ethics as both a matter of head and heart - people can do the right and the wrong things for either reason. For some situations it may be necessary to seek advice from the teachings of sociologists and psychologists. On the subject of business ethics there is a wealth of publications to review. (4)

> *... a matter of head and heart.*

MANAGING TECHNOLOGY
DEPENDENT OPERATIONS

Managing Age and Experience Differences

With respect to age differences, there are the concurrent issues of education, training, and attitude towards the value of experience. From a functional manager's perspective, adoption of new technology and benefits of experience to evaluate its applicability is a troubling task. The young engineers and technicians come to the job fresh with the latest technical tools, be it from a college or trade school. They know of nanotechnology, genenome, and use the latest dynamic programming and simulation techniques or the latest computer options. They handle the laptop features and "surf" the Internet instinctively without asking why. They simply say when you do this, this is what shows up on the liquid crystal display, this is how you transfer these files, rotate this view, etc. They live in a digital world. This of course does not mean that those with some gray in their hair cannot do the same, but it does mean that they have to make a special effort to stay abreast or catch up.

On the other hand, the fascinating advances of the very visible and publicized technologies place into the distant background the core of industry and the experience on which it is based. As examples: efficient production of oil, delivery of clean drinking water, waste disposal, the manufacturing of parts for fuel-efficient cars, and the total quality aspects of the production of food and pharmaceuticals are all vital. Computer literate MBA's are not going to insure that plants will run efficiently or new innovations will come into practice without some experience of simply doing it. The famous Murphy's Law, i.e. anything that can go wrong will, has validity in practice. The young need to recognize this. The whole issue boils down to an aspect of knowledge management. (See Table 5) In many traditional operations this gap between the young technicians and the older experienced personnel is more apparent in the production department and plant operations than anywhere else. The gap must be bridged between the young, fresh-out-of school employees and the senior staff. This must be done through planned training programs. Some of these may require technical training, while others need motivational training.

> *... issue boils down to an aspect of knowledge management.*

MANAGING DIVERSITY – TECHNICAL AND CULTURAL
By Donatas Tijunelis

Table 5: Knowledge Management in Operations

For The Experienced - On the Job:

- To Overcome NIH (not-invented-here) Factor
- To Adopt a Continuous Improvement Philosophy
- To Catch Up and Keep Up with Technology

For The Young and UN-Experienced:

- To Appreciate Total Quality and the Baldrige Award Criteria
- To Get Hands-on Participation in the Operations Processes
- To be Mentored and to Learn of Possible Application Issues

For The Experienced - From Other Jobs:

- To See the Legal and Practical Limits of Technology Transfer
- To Learn Opportunities for Team Building
- To Catch Up and Keep Up with Technology

Professional societies, trade associations and universities offer a variety of technical training courses. At a minimum, these will bring awareness of the gaps and a stimulus for self-study and the honing of skills. Motivational training can be through individual consultants, possibly recommended by the Academy of Management, and courses offered through American Management Association, professional societies, and continuous education programs at various local education institutes. One particular tool that may be helpful to point out any training needs in management of diversity is the Baldrige Award Criteria (5). The implementation of

> *...review the balance of technical, social and business aspects that have an impact on the performance of operations.*

the suggested criteria, by itself, does not guarantee business success. It benchmarks and provides a mature view of a business operation as it is affected by the synergy of its functions. The criteria does force a review of a balance of technical, social, and business aspects that have an impact on the overall performance of operations (see example in Table 6). Since the intent of the criteria is for the self-analysis of a firm with respect to its competitiveness, it should be understood and appreciated by all. However, leadership's significance, the appreciation of strategic planning and its deployment problems, customer relations, and business impact in particular may be lacking in those entering operations organization directly from a purely technical training program. Therefore, its greatest value is to be noticed by the novices in industry.

Table 6: The Baldrige Based Criteria and Relative Values

1. Leadership 12.0%
   Organization leadership & social responsibility

2. Strategic Planning 8.5%
   Strategy development & deployment

3. Customer and Market Focus 8.5%
   Market knowledge & customer satisfaction

4. Information Analysis..9.0%
   Measurement of performance & information management

5. Human Resource Focus..8.5%
   Work systems & employee education, training, & satisfaction

6. Process Management..8.5%
   Product, service, business & support processes

7. Business Results...45.0%
   Customer, financial, market, HR, organizational results
                                                                                                     100.0%

MANAGING DIVERSITY – TECHNICAL AND CULTURAL
By Donatas Tijunelis

A single free copy of the criteria with more extensive explanation and forms can be obtained form the National Institute of Standards and Technology (NIST) *(http://www.quality.nist.gov/)*.

The diversity gap between the young and the "old," if not addressed, can widen and operations management can become more difficult in the future. After all, most new technicians, craftsmen, engineers, and scientists are recruited into operations for their technical skills. Their input is in the advanced techniques that they bring. They get applauded for their technical innovations. They are busy keeping up with technology advances in their specialty. Through a narrow view, management may be reticent in diverting them to broader training. This would be a mistake. It would make the gap widen and diversity management more difficult.

Managing Interdepartmental Differences

Depending on the size and business nature of an enterprise, various aspects or operations can take on formal or informal roles. In a small organization roles tend to overlap while in a larger, more formal organization,

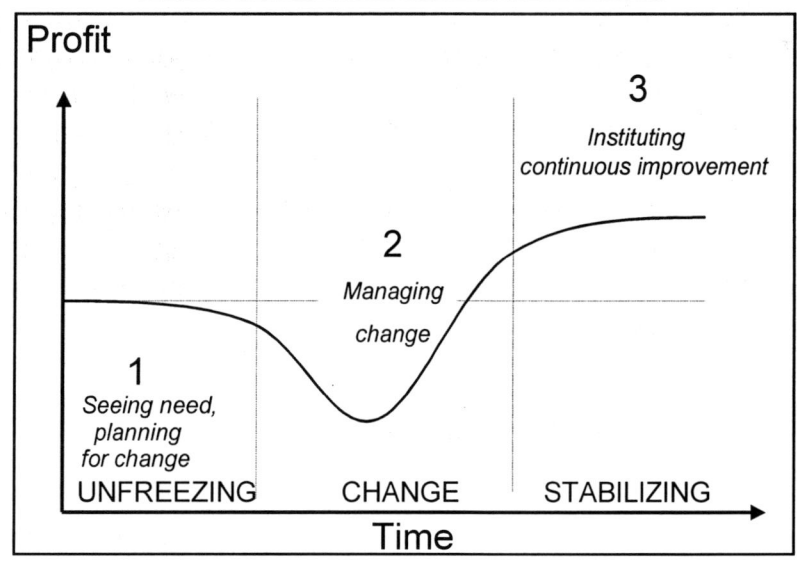

Figure 1: Stages of Change Process

MANAGING TECHNOLOGY DEPENDENT OPERATIONS

they are separated and therefore become a problem to manage. The most obvious differences come about by the inherent mission of various functions. This is most noted under the circumstances of change. Each function may take a different view of the general process of change (Figure 1).

The process starts by unfreezing the status or any negative trend with a plan for change. Then the change needs to be managed for the desired result and finally stabilized with continuous improvement plans. The different reactions to change can perhaps be best illustrated through changes that accompany a product's life-cycle. Consider the following possibility:

> *...manufacturing is likely to be most conservative about changes.*

At the beginning of a product change or new product introduction, it is the R&D personnel that are challenged. Marketing will study the potential and may encourage change. They speculate. However, to prove the feasibility of a new technical product or process feature, R&D has to deliver under the pressure of negative cash flow. In the mean time, production plant process managers are likely to take a dim view. Manufacturing is likely to be most conservative about the proposed product or process changes. Their charge is efficient and stable operation, not radical changes. Changes are likely to bring them headaches, while the recognition for success will be going to sales, marketing, or R&D. This can be seen by tracking the product life cycle features shown in Figure 2. Marketing and Sales will be the first to report good news if the innovation begins to succeed.

The messenger of good news gets the attention and the first rewards. With good fortune, sales will become predictable; process changes stabilize and, in practical cases, enter the state of limited improvements. From the production management's point of view things finally look good. Unfortunately competition by now responds to the challenge with product differentiation or cost competition. At this point marketing comes to R&D for innovation, who in turn are back at the operations manager's desk pleading for some time on the production line to experiment with new ideas. Production efficiency is to be balanced against possible long-range overall results. Squabbles emerge for operations management to handle.

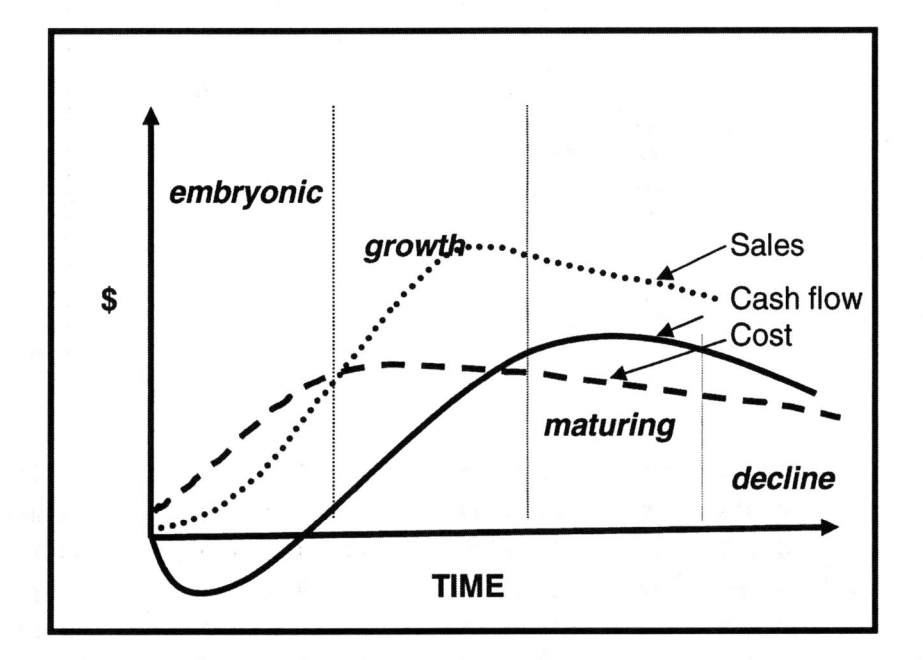

Figure 2: Product Life-Cycle Features

People in their functional roles should not be stereotyped. There are many very forward-looking plant managers as there are conservative marketing directors. However, while not universal, the above scenario in one form or another is common because of the expectations placed by the organizational mission and roles of the people involved. Functional organization "silos" exist, and there needs to be an effort to reduce their effect. Depending on the circumstances and the extent of problems, consider any of the following (Table 7):

Table 7: Addressing Diversity Problems Between Operational Functions

ACTIVITY / INSTRUMENT OPTIONS	FUNCTIONS of CONCERN	PLANNED ACTION Yes	No
Cross-training and mentoring programs			
High-performance, empowered team training			
Reorganization to matrix or project format			
Organizing project teams with diversified representation			
Instituting projects early in the change process			
Communicating strategic plans across functional lines			
Recognition awards for cross-functional contributions			
Oversight/advisor board for cross-functional cooperation			

In larger organizations, the separation of roles or functions are further increased by professional discipline differences of the staff. In a manufacturing operation, the production people may be industrial engineers, while in the engineering department they may be mostly mechanical, while while in R&D they may be chemists and so on. Each has a language of their own, their own preferences and biases. Management training programs that focus on interdisciplinary issues help reduce the distinction and promote synergy. Note, some one-day technical or management seminars, whatever they contribute to technical advancement, do a lot for the appreciation of interdisciplinary issues and mutual respect.

Harmony Paradox and Influential Attention

Management cliques develop at different levels in the organization. They compete and seek to gain recognition by those one or two levels above.

MANAGING DIVERSITY – TECHNICAL AND CULTURAL
By Donatas Tijunelis

While privately disagreeing with management, people are reluctant to voice their opinions in public. Official differences with those in power at upper levels are not what careers are built on. Agreement tends to prevail vertically. Since most management seek harmony among their subordinates, within a given management level "group-think" can be inadvertently encouraged. "Whistle-blowing" becomes discouraged. Excessive cooperation can become self-defeating. In practice, top management is focused on "the bottom line" financial issues. In contrast, the operations staff may be focused on technology without much appreciation for their potential financial impact within the strategic goals. What results is disparity between levels of management. At the start when new concepts are on trial and costs are still small, top management's attention is lacking. When commitment is made and costs become high, top management attention rises, but opportunity for alternative decisions may be lost. One recommendation would be to take some advice form the "One Minute Manager" (6), which encourages managing by walking around, eating lunch with employees at different levels and departmens in the organization, and asking questions outside the formal reporting lines.

> *...cliques develop at different levels in the organization...*

For the technical person in the process of product innovation, it is important to recognize management's span of attention. As companies release new products, the need may be obvious to some concerned with managing projects already at hand and not so obvious to others. It seems paradoxical that when costs are small, significant consequences are ignored which may affect large future cost. When costs become large, attention goes to incremental cost control (see Figure 3). This subject is further discussed in the chapter dealing with strategic management.

MANAGING TECHNOLOGY
DEPENDENT OPERATIONS

Figure 3: An Example of Attention and Workload Cycles by Function

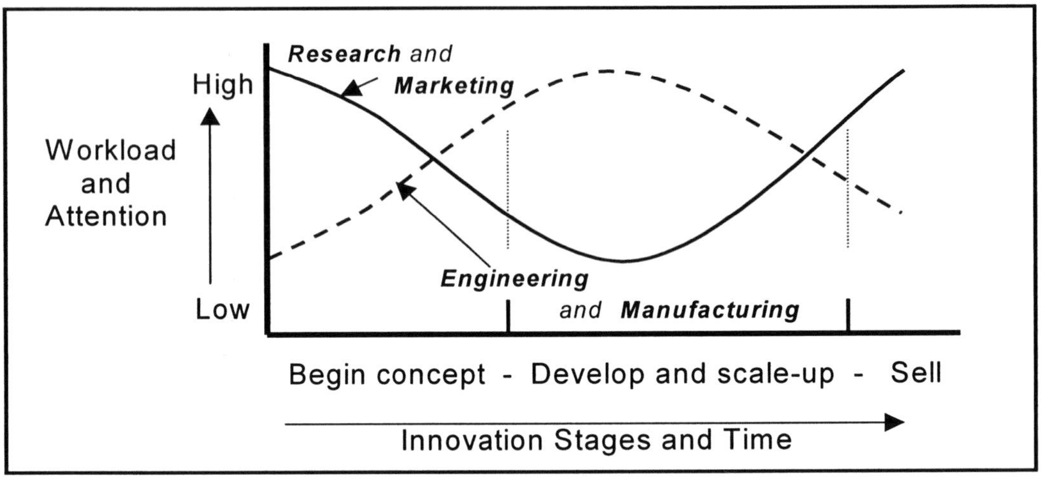

[Figure 3: Graph showing Workload and Attention (Low to High) on the Y-axis versus Innovation Stages and Time (Begin concept - Develop and scale-up - Sell) on the X-axis. Research and Marketing curve starts high, dips in the middle, and rises again. Engineering and Manufacturing curve starts low, peaks in the middle, and declines.]

It is up to operations management leadership to recognize the significance and respond accordingly. Management leaders have to recognize the motivational bias at various levels. Avoid the assumption that one discipline or organizational function will look out for the needs of another. What is normally considered, a positive attribute, core competency, can be a negative. It can lead to overconfidence. Core rigidity may come into play. Programs, projects, and individuals in a fast changing environment may try to sustain their own competitive advantage within an organization, even hope to expand it, while the scope of their credibility diminishes. It is typically game of internal politics. A pure administrator or technical micro-manager may not recognize this fact. It must be the responsibility of the operations leader. One possible tool is to ask each functional unit to periodically do a self-analysis as a small survey of their strengths and weaknesses, opportunities, and threats (SWOT). When reviewed it can point out the inappropriateness of their workload and the possible need of higher management's attention.

Core rigidity may come into play.

270

MANAGING DIVERSITY – TECHNICAL AND CULTURAL
By Donatas Tijunelis

International Management Issues

When managing a technical function across borders, be it as a technical organization or individual projects, additional complexity arises. Individual difference in the form of national cultural traits between the people involved is one obvious factor. In addition, there can be a list of other issues to consider. These are not isolated, but are interdependent. Some, while appearing to be technical, are the result of social and political motives or the reverse, i.e. the root cause of the social/political factors may be economic. Technology application, as would be encountered in managing operations, is not inherently independent or objective as science is. It depends on the cooperation of people, their attitudes, and the organizational system in place. In teaching international management to business managers the following topics are typically included:

> *... technology across borders... depends on national culture...*

1. The acceptance or rejection of a trend toward globalization
2. Factors of social and cultural differences
3. Political economy and national competitive advantages
4. Trade regulations, theories and practices
5. Monetary policy and currency exchange fluctuations
6. Foreign direct investment options and trends
7. Competitive strategy and operational issues

The above list of topics should be reviewed when examining possible challenges in managing technology or technology transfer situations across boarders.

MANAGING TECHNOLOGY DEPENDENT OPERATIONS

> *Gobalization is resented where national pride is threatened...*

Globalization is essentially a given. The Internet has in effect made nations less isolated and more interdependent. Like it or not, the world has gotten smaller. The supply chain has become shorter. Access for nearly anything the producer needs can be outsourced and what the consumer wants can come from anywhere, if there is the money to pay for it. Some say outsourcing causes large businesses to dominate and exploit cheap labor, making the rich richer and keeping the poor dependent. Others argue that globalization helps technology transfer and without it the poor would be even poorer.

Gobalization also causes cultural homogenization. With transfer of technology comes transfer of consumer goods, assimilation of tastes, lifestyles and a change in local culture. That is often resented where national pride is threatened. It causes emotional distress as seen at the site of the World Trade Organization meetings.

As a manager of technical programs that crosses national boarders, requiring cooperation on both sides, it may be good to ask a few questions. Be prepared. For example, if globalization is a "hot" and controversial topic in the region, get updated on the issues and be sure that your technical staff is updated. Look up related current publications – perhaps in *The Economist, The International Herald Tribune, The Wall Street Journal,* or the local papers. Make some effort to feel out the key participants on both sides of the border regarding their attitudes toward the issues. Be sure that your technical staff assigned to the international projects recognizes the complications that can arise due to international political factors and are prepared to handle them accordingly. When in doubt, for a situation where security may be an issue, the US embassy is a good place to turn to for advice.

Globalization is variably defined. However for operations management it can be condensed to the shift towards a more integrated and interdependent world economy in two principal ways:

MANAGING DIVERSITY – TECHNICAL AND CULTURAL
By Donatas Tijunelis

- **The globalization of <u>markets</u>** ………"make it here and sell it over there"
- **The globalization of <u>production</u>** …"make it over there and sell it anywhere"

Staying with the topic of operations, the focus is on globalization of production. In one form or another, foreign production strategy decisions involve foreign direct investment (FDI). These can be examined by the following sequence:

1. <u>Is the product exportable at low shipping costs, minimal trade barriers?</u>
- If yes then, simply make it here and sell there.
- If not, foreign production option comes up.
2. <u>Can the production technology be applied under foreign conditions?</u>
- If yes, besides capital needs, issues of technology transfer come up.
- If not, and the technology is protected, then licensing may be an option.
3. <u>Can the quality and brand reputation be protected?</u>
- If yes, license
- If not, consider training of foreign operations.

Smaller companies take an opportunistic view. They organize themselves to fit the circumstances. If they can make it here and sell over there they will do so, primarily adjusting in the selection or training of their technical staff. If the customer overseas is big, it is likely that conformance to ISO standard requirements will be needed. Then an appropriate training program of their operations staff will be required. In a small company, the role of foreign-trade consultants become important. Their comprehension of the foreign trade practices may or may not need scrutiny, but their understanding of the client's technological strengths and weaknesses, as

MANAGING TECHNOLOGY DEPENDENT OPERATIONS

well as technology transfer issues should be tested. A good practice is to have a technician accompany the consultant during any more serious foreign contact.

> *...foreign-trade consultant should be tested...*

Large companies develop an organizational strategy that accommodates international business needs. Recognized experts in international management say that their challenge is to balance an organization to be locally responsive, while at the same time seeking to exploit cost reduction opportunities through low cost labor or capacity utilization worldwide. Such large companies respond with organizational strategies that fall into four categories (7):

- GLOBAL - domestically-centralized technical operations exploiting experience with standardized global products seeking foreign sales and distribution. (Advanced electronic firms like INTEL, Texas Instruments)

- INTERNATIONAL – firms with domestic R&D, high core competency establishing manufacturing operations and marketing in the countries they enter. (Some consider Procter & Gamble, Microsoft, IBM in that category)

- MULTIDOMESTIC - firms customizing their product offerings, organized into totally independent operations in each foreign country. (A good example may be Unilever)

- TRANSNATIONAL – Few firms with unique matrix management culture that tries to be flexible with intent to learn and capitalize on diverse centers of excellence. (Some see Caterpillar and Nestle seeking such strategy)

In the context of globalization, operations management has to take on a number of roles. While these are quite distinct for large multinational corporations, they somewhat overlap in small companies. In large corporations the difference in roles relate to organizational responsibilities as follows (8):

- *Corporate Head*: Leader-talent scout-developer
- *Business Manager*: Strategist- architect- coordinator
- *Country Manager*: Sensor- builder-contributor
- *Functional Manager*: Scanner- cross-pollinator-champion

As corporate leader - a president/CEO/business owner - the role is one of international organization building. They guide policy, staff at the executive level, and affect corporate culture. Their challenge is to bypass some protocol and get down to relate to the people at all levels everywhere. Foreign organizational bureaucracy will be a barrier. Some of it is the result of cultural differences and some simply can be a form of corruption. Some like to refer to the saying "when in Rome do as the Romans do." However, in the case of corruption, there is no business like good and honest business. Avoid corruption. To do so, one has to learn to recognize it. The main remedy is travel, repeated exposure, and perseverance. Special executive cultural diversity training programs have been known to be helpful.

> *... remedy is travel, repeated exposure and perseverance.*

The business unit managers are concerned with the bottom line of their operation that can span several countries. They need to coordinate functional units across the foreign culture difference. A strategy for competitive advantage must acknowledge cultural preferences for products, economic condition differences and nationalism within the ranks of the organization across borders.

A country manager is truly the local operations manager. The role is the same as the general manager of a domestic division or a smaller company, except that the profits go to a distant corporate headquarters or a distant

MANAGING TECHNOLOGY DEPENDENT OPERATIONS

owner. For the purpose of selecting a resourceful builder that understands the local situation, a native of the county is advisable. However, here the challenge is that of loyalties - loyalty to the country versus that of the corporation.

In the case of the functional - production, engineering, R&D manager - the challenge is to be objective. Because of distances and national pride, cultural factors and the "not-invented-here" or the "NIH" factor can show up. Regardless whether it is technical operations personnel of a large corporation or the technician representing the small firm, cross-cultural training may be needed because of the minimal exposure that technical discipline provides in political and social education. Technology transfer is difficult within a domestic operation between R&D and production; it is much more so on the international scale. Consider the probabilities of innovation to proceed from concept to commercialization:

Assume that technology itself has a 30% chance of reaching the stage of production scale-up domestically. That same technology developed across borders could have only 20% chance for the reasons discussed above. Going further, if the scale-up to production on the first attempt is estimated to have a 50% probability domestically, it may be reasonable to think that additional technology transfer issues would reduce it to 40% in a foreign country. Finally, sales people, with their natural optimism, may assume a probable domestic sales effort for a new product in a known market to be 75%. For the new product in a yet unknown market, it may be 50%. In total, the probability of successful commercialization domestically would be 0.3x0.5x0.75 or ~ 11%, while in the global arena 0.2x0.4x0.5 or only ~ 4%. The hypothetical example may be overly generous, but hopefully makes a point: if innovation is challenged with many hurdles domestically, it will have much greater hurdles to handle

> *Technology transfer is difficult within a domestic operations; it is much more on the international scale.*

internationally. Careful planning and possible cross-cultural training to recognize and to effectively manage cultural diversity are in order.

Social factors are important to take into account. The question of proper behavior to gain cooperation and respect internationally is well highlighted in the title of in a popular cross-cultural training book *Kiss, Bow, or Shake Hands* (9). The choice to adhere to foreign rules of conduct is important. Cultural differences, as they relate to the values in the workplace, have some subtle features that need attention when examining potential international management situations (Table 8) (10). Each cross-border situation should be examined separately. The results of any available study for the specific industry and country involved should not be carried over into another situation. Once again, it is unfair to generalize and stereotype. In many cases the best way to assess the potential impact of cultural differences is by traveling and experiencing them first-hand. However, use can be made of the categories of work-related national cultural characteristics to help foresee and avoid possible insensitivity and conflict. In all cases, each person must first be treated as an individual whose behavior may or may not exhibit any traits broadly attributed to the society of which they are a part.

Table 8: Cultural Differences and Values in the Workplace

Work-Related Values	Feature of Work Roles at The National Level (Possible questions to consider)
Power distance:	Acceptance of unequal physical and intellectual differences (Question: Is title/degree important? How will disabilities, age be treated?)
Individualism vs collectivism:	Relationships to families, peers, independence and competitiveness (Question: How to choose team leadership, will nepotism be a problem?)
Uncertainty avoidance:	Acceptance of ambiguous situations (Question: Will decisions come slowly? Win with patience or quick wit?)
Masculinity vs Femininity:	Relationships between gender and work roles (Question: Is it a male dominated society? How will a woman be respected?)

MANAGING TECHNOLOGY DEPENDENT OPERATIONS

As an example, citizens of US and Australia tend to have very high regard for individualism and want to take charge, compete. The Japanese and French rank high with respect to uncertainty avoidance, requiring patience in negotiations. Latin Americans in Panama or Mexico have been noted as high in the power distance dimension - valuing titles and organizational structure. Sweden is low in the masculinity vs femininity dimension, implying little distinction between the roles of men and women in their workplace roles.

The above examples are given to ponder for the more common situation of managing technology across borders. Of course, beyond these there are the extremes in religious beliefs and family and personal experiences. The impact of religion and purely political issues can be overriding. Who eats what food, abstains from which stimulants, or which country is in the state of war with another can be critical and a separate topic in itself.

It may be useful to both the technical staff and the general management to take account of the national competitive factors when entering with operations into a foreign country. What can be expected from the local environment with respect to its impact on the technical and business objectives? Consider Table 9 adapted in part from Michael Porter's concept (11) of national competitive advantage. Ask appropriate questions with regard to possible challenges to meet technical objectives. Then prepare accordingly. Check to see that wrong assumptions are not being made with regard to technology transfer and that the diversity of circumstances is considered.

Table 9: Features of National Economy for Competitive Operations

National Feature	Description	Possible implication for Operations technology
Local Endowments	The economic environment locally	Are climate, local skills, labor pool, natural resources OK? Staff housing conditions?
Local Demand	The purchasing power, demand locally	Will high local demand lead to local control or the reverse create lack of local interests and changes?
Infrastructure	The built-in support services	Are there adequate support industries? Are there industry associations? Standards?
Strategy	Has the local government a supportive strategy	Do the local authorities support foreign investment? Regulations OK? Is intellectual property protection OK?

In summary: It is likely that in the future technology advances will challenge operations management by the increasing diversity of skills for various age groups. Continuous learning will become essential. Knowledge management will involve organizational behavior that complements technical goals and organizational development that considers job enrichment and ethics as part of a global strategy. To benefit from globalization and the learning to manage across borders will require of the technically trained individual to appreciate other cultures, other economic factors to capitalize on technological advantages. Unrestricted travel, Internet data sources, and Web-based virtual project teams are the most immediate tools for effective management of technology in a global environment.

References

1. S. Altman, E.Valenzi, R.Hodgetts, *Organizational Behavior: Theory and Practice*, Academic Press, Inc., Orlando, 1985, p78
2. O. Kroeger and J. Thuesen, *Type Talk at Work*, Tilden Press, 1992
3. S.R. Covey, *The & Habits of Highly Effective People*, A Fireside Book, Simon & Schuster, New York, 1989
4. F.Neil Brady, *Ethical Managing, Rules and Results*, MacMillan Publ. Co., New York, 1990.
5. Baldrige National Quality Program, National Institute of Standards and technology (NIST), Route 270 & Quince Orchard Road, Administration Building, Room A635, 100 Bureau Drive, Stop 1020 Gaithersburg, MD 20899 or e-mail: *map@nist.gov.*
6. Kenneth Blanchard and Spencer Johnson, *The One Minute Manager*, Berkley Books, New York, 1983.
7. C.A. Bartlett and S. Goshal, *Managing Across Borders*, Harvard business School Press, Boston, 1989.
8. Chreistopher A. Bartlett and Sumatra Goshal, "What is a Global Manager?", *Harvard business Review*, Harvard Business School, Boston, September-October, 1992.
9. T. Morrison et al, *"Kiss, Bow, or Shake Hands"How to Do Business in Sixty Countries*, Adams Media, Corp., Holbrook, Mass, 1994.
10. Based on G. Hofstede, *Culture's Consequences*, Sage Publ., 1980
11. Michael E. Poryer, "The Competitive Advantage of Nations" *Harvard Business Review*, Harvard Business School, Boston, March-April, 1990.

Chapter 9, Lead-in:

STRATEGIC ISSUES IN MANAGING TECHNOLOGY

Whether it is a small business competing in a fragmented market or a large muli-national enterprise, it is very likely that more than one aspect of competitive environment is "on the table," more than one project is on the horizon. Strategic issues are likely to be viewed differently by the general manager and technical specialist. Yet technology, more often than not, is the basis of competitive advantage. Strategic thinking and technology management goes hand-in-hand. This chapter addresses the need for the generalist to appreciate how the technical specialist may prioritize his or her work. At the same time, it is intended to show the technical specialist what the business manger considers as strategic and critical.

To be on the same page, the very basics of both strategic management and technology management, in terms of project portfolio prioritization, are discussed in this chapter. Tips and tools are described that could help "sell" a project. Suggestions are given for simple ways to estimate the value of a project and for a general manager to prioritize them among several alternatives.

The underlying message is to keep technical program strategy simple but effectively used. Project portfolio is to be prioritized according to probable economic impact using metrics that are most suitable to the situation at hand.

CHAPTER 9

STRATEGIC ISSUES IN MANAGING TECHNOLOGY

Donatas Tijunelis

It is obvious that both the worldwide economic environment and technology are changing at a very rapid pace. Businesses have to deal with it. Management is asked to be generalists to deal with the economy and specialists to stay on top of technology, both at the same time. It is not an easy task. For competitive reasons, corporations are restructuring, downsizing, and developing flexible organizations. As innovation accelerates, project based organization forms are expanding. In response, emphasis has been placed on MBA curriculum aimed to stimulate entrepreneurship. Marketing, human resource, communications, and financial management have gotten a great deal of attention. The management of technology in the form which is needed for operations, however, has often been left to on-the-job training. Consequently, (operations) management now faces the conflicting problems of a demand for quality, higher productivity and the never-ending growth in product innovation with little preparation. A technology-trained operations manager needs to appreciate business impact, just as the generalist should understand the basic elements of technology management. While strategic management at a corporate and business level is a popular subject covered in most basic

> *...project based organization forms are expanding.*

management training programs, its interrelationship with project management is a subset that is often left unrecognized.

Strategy implementation for changing technology ends up in one form or another as the management of projects, yet how one defines a project will vary. For instance, the building of a factory is a project for which plans are being laid out in a very detail and formal manner. The selection and the installation of a new programmable controller on the take-up reel of a plastic film extruder may be much less formal, yet still a project. Then there are the more publicized projects: R&D programs to develop a new computer chip or a genetically engineered species of corn. It all ends up with someone making choices of what to do, how long it will take, who is best to handle it, what resources are needed and what is its potential value. Of course ultimately, from the business point of view, who needs it and why do it? Yet it is not unusual to find that strategy developed to answer these questions is unclear to even a business manager who came up the ranks without exposure to technology. The concept of the goal and the value of the effort may differ between the technician and the general manager, but it should not. The general manager and the project leaders of the technical staff have to have at least a feel for how business strategy evolves – why certain decisions are made and how to adjust accordingly.

> *...the generalist should understand the basic elements of technology management.*

What is Strategy in Business Management?

To begin, what is strategy? Strategy can simply be defined as the positioning for a sustainable advantage. At a top corporate management level it means making choices of what industry to stay in, get in, or get out of. The strategy development process starts out as a mission for the corporation and leads to a vision of how best to compete at the business level product by product. It then further evolves into an implementation strategy for each

MANAGING TECHNOLOGY DEPENDENT OPERATIONS

function of the organization. Ultimately it ends up as a plan for action for a particular project or a program for a group of projects to be evaluated, prioritized and implemented. This sounds very top-down, but in a dynamic business world it is becoming an iterative process with constant review and revision. This should become clear through the following discussions.

> *...the positioning for a sustainable advantage.*

A strategy for competitive advantage needs business smarts. For its implementation, it needs engineers and technicians to have operational know-how. For a technology dependent operation, input from the engineers and the technicians can provide a key component of competitive advantage. There is little doubt about that. But who should lead, who should follow? Should business strategy be built on the availability of technology or should technology be adopted or developed to fit a business strategy. Sounds a bit like "which came first the chicken or the egg?" In the past, as a practical matter much of corporate, business-unit, and functional operations strategy was top-down. When technology was simpler, this made some sense. The top executives had the broader view. They set the mission. Through the business managers they handed down to the technical staff their objectives. It could be done relatively effectively because technology was more familiar, changed slower. The technician and engineer had fewer choices, fewer options to propose. The business manager could evaluate its commercial value without much help. There was time for management to understand technology and its future opportunities. Technology moved slower then. It is not the case now. US Patent applications in 1900 were somewhere around 40,000, in 1950 around 60,000, but from then on it rose exponentially. In 2001 there were over 340,000 of which only two of the top ten corporations granted US patents were US companies (IBM and Micron Technology). Strategic

> *...do managers of technology value business prospects? Do they care?*

management became complex. Both the technician and project manager in the operations are needed to contribute to strategy development. But do the functional managers see the technical features in their business prospects? Do they care? They should.

Why Projects and Development Programs Need Strategic Thinking?

Operations management has two features. One is effectively maintaining steady state and the other is managing change in line with market needs and availability of resources. To manage change the trend is to rely on project management techniques. Organizing by key projects allows the start and finish of strategic effort in units that can be managed more easily. It helps to be more flexible - staffing as needed and outsourcing where appropriate.

Project management is extensively and almost exclusively used in many types of organization where project are the normal way of conducting business, i.e consulting firms, construction firms, contract R&D organizations, governmental organizations, etc. In these organizations the normal mode of doing business is via projects. The organization proposes a project to a client or customer and after customer approval, this project plan is effectively a legal document between two organizations. In these situations, the projects are laid out in great detail with costs and times built into the contract. Any deviations from the project plan result in a contract default with lawyers and accountants immediately involved. In these instances the project plan has become a contractual document and can only be modified via formal legal actions. The project manager, whether or not that person was involved in preparing the proposal, is responsible for performing based on the proposal. Any deviations are considered a failure. In the chapter, this is referred to as traditional project management.

The coverage within this chapter focuses on project management for projects that are "within the family." This is the normal situation with all organizations that are not selling projects externally. In such situations the customer is internal. It may be a different department, different site, etc, but it is still part of "the family." The organization conducting the projects and the customer are both working for the overall good of the overall organization. The project manager not only can but should be involved in managing for the overall good of the organization. "Within the family" the interactions between management and project management become a cooperative activity for the benefits of the overall organization. It is this "within-the-family" view of the interaction between management and project that is discussed it this chapter.

> *...A large part of a traditional project management task is mechanistic.*

For traditionally project management attention has been relatively narrow. It was internally oriented to optimize resource allocation, develop a work breakdown schedule, and prepare progress and cost control reports. In this context it takes on a predominantly administrative role. A large part of the task is mechanistic. Recent changes in the available technology and global economy have the following two consequences. One, many of the mechanistic tasks, like data management and schedule maintenance, can be delegated to the computer, thereby freeing up time for higher level decisions. The other is the opportunity for management to reorganize, to search out virtual resources through the Internet, to be more flexible with fewer people on staff – to be more entrepreneurial. More outsourcing in a global economy leads to a broader range of stakeholders with complex interdependencies. At the same time, a trend to reduce management leads to less supervision of the workers - there is more autonomy at all levels of the organization. Consequently everyone should understand the overall strategy of the organization - know the intended direction and adjust accordingly as the situation changes. Today, project management has become a key part of the overall organization and the project manager has broader responsibilities (Figure 1).

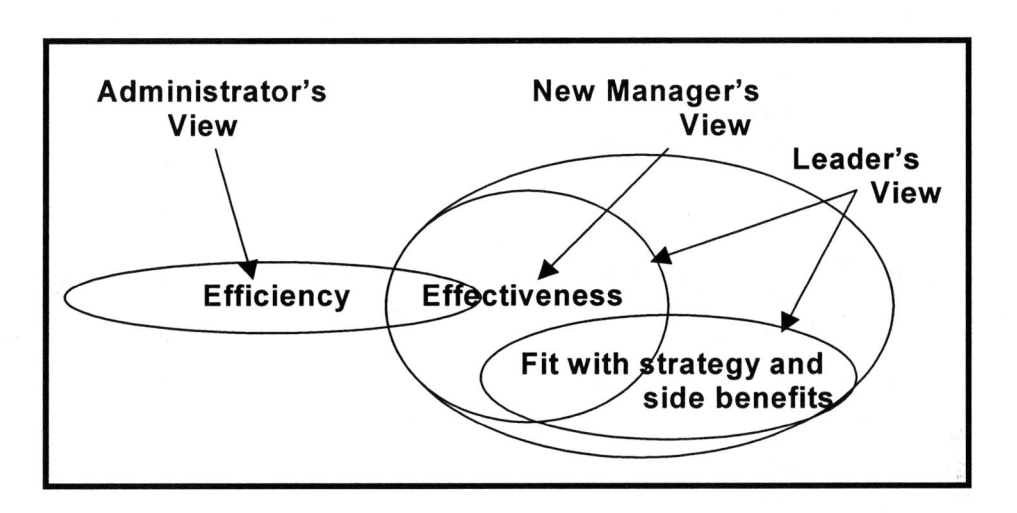

STRATEGIC ISSUES IN MANAGING TECHNOLOGY
By Donatas Tijunelis

Figure 1. Dominant Views of a Project

In private industry, project managers in a leadership role need to consider when projects can be pooled into a program targeting the same general business goal and when they need to be isolated. Some exploratory projects may need to be isolated for the sake of objectivity and confidentiality. On the other hand, others can and should be combined into a program of related projects to bring administrative efficiency - more effective utilization of skills and physical resources. Some refer to these programs as super-projects. The direct or indirect dependencies of a project on other projects within a program are managed to avoid duplication of efforts, conflict, and to strive for synergistic results.

> *...strive for synergistic results*

Each project manager should know the strategic intent of their project and with that perspective, structure project plans for the broadest and most effective impact. As much as possible, projects should be anticipated in advance through some form of strategic planning or at least through the process of strategic thinking. The latter can be just a habit of proposing and/ or reviewing the goals with regard to internal resources and external competitive conditions for the probability of commercial consequences in the future. The trick is to be insistent that it is done in each case and yet to be flexible in the form. This requires some understanding of the strategic planning process and project evaluation techniques and so it leads into the subject of the following sections.

Who Does What in Strategy Development?

> *...compete for the future – think strategically...*

Corporations have a strategy development process that defines the basic procedure and who does what. The responsibilities for input issues and the type of decisions generated from these vary depending on the level in the organization.

At the corporate management level of a multi-faceted corporation the principal strategic decision is to clarify in what industry they want to be, to define who they are and what their mission is. For example, are they to be identified with auto parts manufacturing, investment banking, or a conglomeration of a number of businesses in various industries. Are they a benevolent not-for-profit or a profit-seeking organization? What is their mission and vision for the future? At this level public corporations compete for investors attention and stock market effects.

- Key issue: Who are they?

At the business management level of a particular product or service, the challenge is how to compete within the industry already specified by the corporate strategy.

- Key issue: How to compete?

At the functional management level of a business, the discussion is about the strategies for the effective performance of various sectors of the organization that create the value on which it can compete within the industry.

STRATEGIC ISSUES IN MANAGING TECHNOLOGY
By Donatas Tijunelis

Here the conversation is about a strategy for sales, distribution, marketing, manufacturing, product development, accounting, or support services

- Key issue: How to best implement within each operating function the business plan for a product or service of the corporation?

<u>At the project management level</u> in a volatile business environment, most of the competitive advantage is created. Any change in corporate strategy ultimately depends on the implementation capability and the planning and execution of projects. (Figure 2)

- Key issue:
    1. How to prioritize and support project activities?
    2. How to institute the project leadership role?

It should be relatively obvious that every aspect of an organization should have a strategy that fits and reflects the strategy of the corporation. Every functioning unit should complement the mission and strive to implement the vision set by the corporate and business leadership. At the same time the vision of the corporation can not be established in absence of an understanding of what is, in fact, possible to accomplish with the resources at hand and the environment at the working level - the project. Project strategy can not be established without knowing where the corporation is heading, nor can a corporate vision of its future be realistic without feedback from the project leadership of its implementation potential.

> *Project strategy cannot be established without knowing where the corporation is heading,*

MANAGING TECHNOLOGY
DEPENDENT OPERATIONS

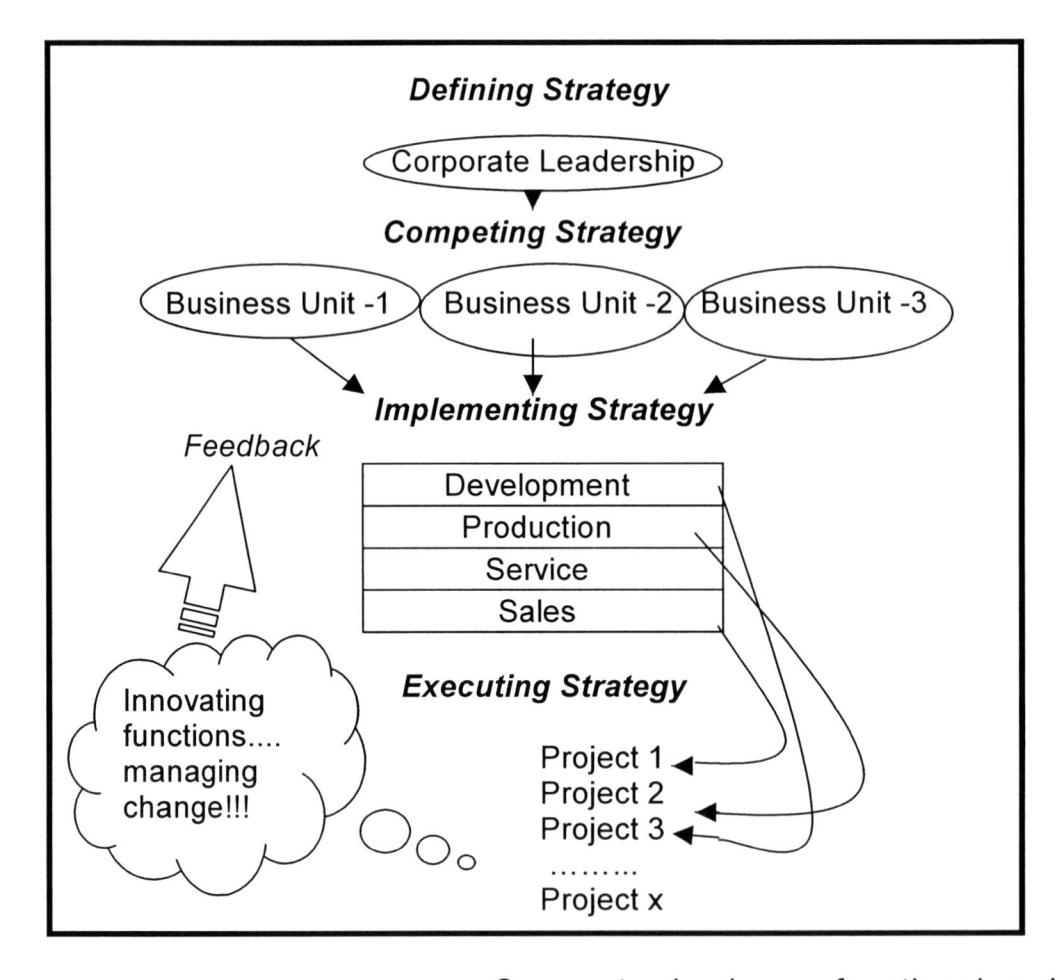

Figure 2: Strategy Roles by Organization Levels in Industry

Defining Strategy
- Corporate Leadership

Competing Strategy
- Business Unit -1
- Business Unit -2
- Business Unit -3

Implementing Strategy

Feedback

| Development |
| Production |
| Service |
| Sales |

Executing Strategy
- Project 1
- Project 2
- Project 3
- ………
- Project x

Innovating functions.... managing change!!!

...compete for the future – think strategic...

Corporate, business, functional and project strategies are interlinked and depend on each other. As the pace of technological change accelerates and as companies grow rapidly and become increasingly decentralized, the ability of front line employees to execute a company's strategy without close central oversight becomes vital.

STRATEGIC ISSUES IN MANAGING TECHNOLOGY
By Donatas Tijunelis

To understand the strategic role of projects in the process of innovation it is necessary to understand what is innovation. In industrial terms, innovation is the process of generating a new concept and converting it into a commercially viable product, process, or service. It is the result of creative thinking, development, and a methodical process that goes through a series of stages. It must result in an application in public use. Projects are created at any step in the innovation process. An exploratory project to look for opportunities or to conduct research and to create a prototype could be the start of the process. The start-up of any new venture is surely a project as much as an effort at quality assurance or an initial product promotion and sales plan. In the innovation process, any one of these steps can be a single project or a program composed of a cluster of projects. A new product and process development program is most commonly a cluster of strategic projects.

> *Projects are created at any step in the innovation process.*

The traditional sequence of projects for new product development was to lead process development. Commercial process development would not start until a product was market tested. This has, however, changed into parallel-development of both the product and its process. The benefit can be readily appreciated. Doing more than one thing at the same time saves time, however with the associated risk of increased cost in the event of a failure.

A funnel (Figure 3) can illustrate the probabilistic process through which the screening of any idea has to go through to reach a project approval and implementation stage. Similar funnel models are sometimes referred to as a stage-gate process with formally defined input, review, and approval procedures.

The IDEO Company, a very innovative and successful design firm in the US, suggests that to "fail early and often is good"(1). It exposes unfavorable options. In doing so the danger of early commitment when attention is lax is minimized. Toyota in the 1990's pushed its product developments faster through a similar approach, which they called "front-end loading." It requires faster experimentation, more projects at the outset of a program to be screened for further development (2). This implies more emphasis on data, early evaluation and quick decisions. However, now is the era of information technology. The ultimate technology for rapid experimentation might turn out to be the Internet, which is already transforming countless users into innovators.

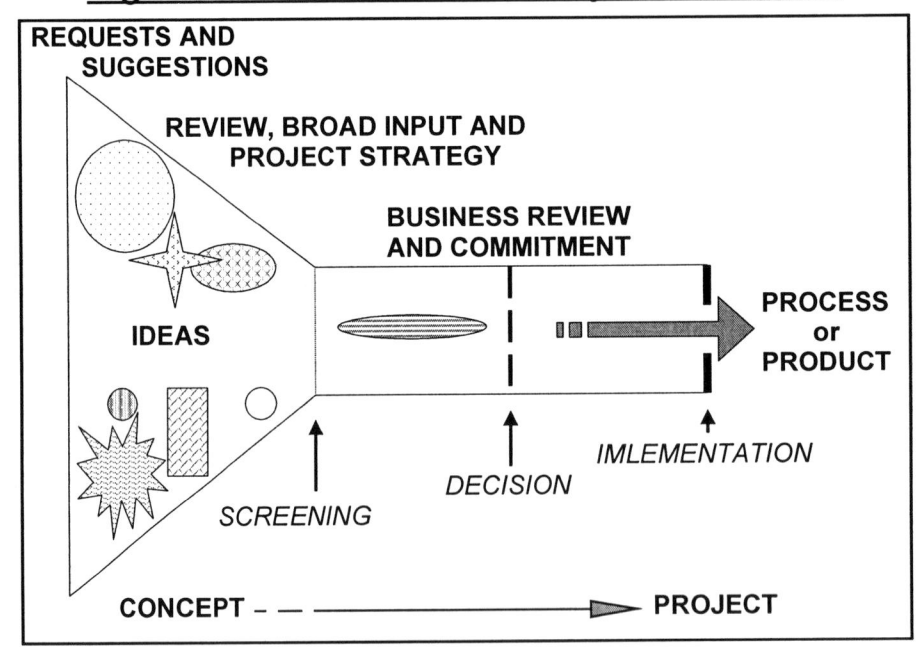

Figure 3: The Funnel of Project Evolution

Strategy Development Basics

The first element of strategy development is identifying the stakeholders. The stakeholders are the owners, stockholders, employees, customers, suppliers, and the public at large. While some may be affected directly and others indirectly, they all have a stake in the results of strategic decisions. Once the stakeholder issue has been considered, strategy development can proceed according to the classic strategic planning paradigm (Figure 4).

Figure 4: Strategic Planing Paradigm

Mission, Vision, Goals
⬇
Analysis of Situation
Internal and External
(SWOT)
⬇
Strategies:
Corporate - Business - Functional - Project
⬇
Implementation Plans

The sequence follows to define the mission, to develop a vision with its goals, to carry out analysis of internal and external conditions, to formulate a strategy and finally a plan of action with specific implementation assignments. Part of a good strategic plan is a brief executive summary bringing up the very key points. Its brevity would insure that it would be read and understood by all stakeholders. A planned follow-up review and revision is essential to keep the plan current.

A mission statement essentially tells the purpose of an organization. The overall mission of a corporation must be clear for effective planning by the business units, the functional departments, and all programs and projects. At the project level, the corporate mission is fixed and defined by upper management. Except in rare cases, project managers do not have much input into the mission of an enterprise.

MANAGING TECHNOLOGY DEPENDENT OPERATIONS

The vision statement of an enterprise describes essentially how it sees itself in the future. It points out any needed direction for change and the major challenges and goals. Traditionally, the vision and its goals were exclusively top management's domain. It often ended up to be an extrapolation of past performance. However, conditions are changing. In the technologically dynamic, knowledge-based environment a project manager may have more to say about the future than does the top executive. Therefore corporate vision has become, in some sense, more reciprocal – it may come from the top, rise as a response to lower level problems, or be a relatively open in need for new ideas. Outside consultants are often used for this purpose. Their objectivity, however, may be suspect. Their allegiance may be biased, keeping in mind that they may favor the perceived interests of those who hire them.

> *... vision of higher-level executives and project managers must not conflict.*

The input for a vision statement can come from an extrapolation of past performance or an assessment of desirable future scenarios. It is a good idea to combine both. Based on hard facts, extrapolate from the present forward through group brainstorming techniques speculating the future and create several probable and desired vision alternatives, then work back to see which of these have a close match with the extrapolations. In effect, the standard approach helps to develop a strategy going from the present to the future, while the other starts with the future and works backward. There is a bias in both and reality probably lies somewhere in between. The key is to avoid getting bogged down by details. It is important to think strategically - to recognize the present, but focus on the future. The goal here is to cover the present and future high priority issues as quickly and as often as possible – evolve a habit of strategic thinking. Review and repeat the cycle frequently. By considering several future vision options and a plan to cover each, the organization will be best prepared to adjust to the fast paced changing economic conditions. The vision of higher-level executives and project managers must not conflict. A project manager should have an appreciation of how and why a given vision was chosen.

A key to effective planning is the analysis of internal circumstances and the external environment that is affecting or will potentially affect the organization and, of course, its projects. This involves an impartial review of the strengths, weaknesses, opportunities and threats. In short, it is often referred to by the acronym of SWOT analysis.

S   **W**   **O**   **T**
Strengths   Weaknesses   Opportunities   Threats

It is advisable that the input of strengths and weaknesses be obtained from more than one source. At this stage of the strategic planning paradigm the project managers can make a direct and very important input. For example, the "customers" of a project may be asked what they consider a risk and why they would chose to sponsor the project. On the other hand, if the project is made up of a team of a variety of participants, a quick survey of their concerns and confidence factors should be collected. If a project has several departments or professional disciplines involved, their views should also be collected. The purpose is to be as objective as possible.

For internal analysis of strengths and weaknesses, recognize that some strengths may become weaknesses. It depends what competency is considered relevant. Technology has become an inevitable part of core competency. The question is "where is the organization's technology in the change process?" Is it just right, far behind, or reaching out too far and thus highly speculative. Core competence can become core rigidity, subject to internal bias. Again, objective assessment is crucial. Outside help for benchmarking can be helpful. Core competency is composed of physical resources as well as capabilities stemming from people, skills, and intellectual property. While physical things are easy to define, the rest are not. The strengths or weaknesses depend on how well operations converts competencies into marketable value. Operations can generate competitive advantage through efficiency, quality, innovation, and response to customer needs. Each of these is a part of strategic plan and therefore can be the project goal in operations management (Figure 5). This topic is covered extensively by Hill and Jones (3).

Figure 5: Sources of Strengths and Weaknesses of Operations Management

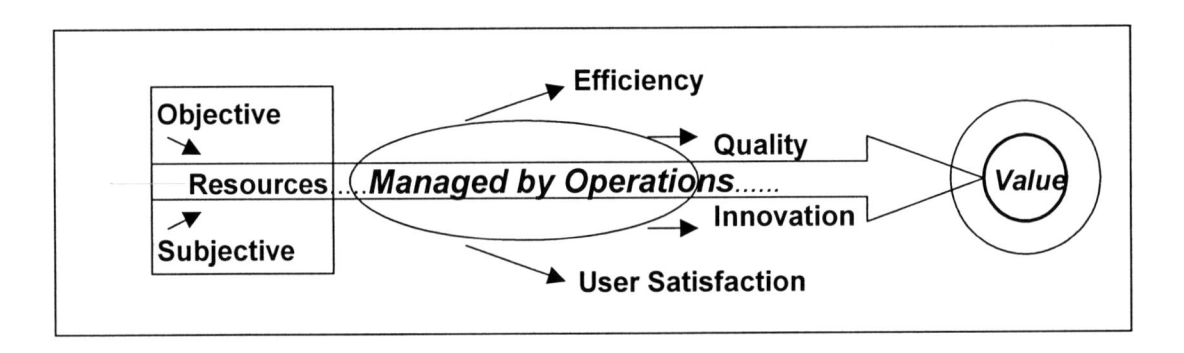

For external threats and opportunities the project leaders should analyze the environment of the project or the program. For an external analysis of opportunities and threats two issues have to be kept in mind. One is the impact of the globalization of the economy. The other is access to vast sources or information and direct communications via the Internet.

The end result is a strategy statement and an implementation plan. The strategy statement presents a plan of action. However, it is meaningful only to the extent that organizational functions can implement it. Operation's programs and projects must be achievable with the available resources and skills. This is where the business plans of any corporation are almost totally dependent on good project planning strategy.

STRATEGIC ISSUES IN MANAGING TECHNOLOGY
By Donatas Tijunelis

Once all that is done, new projects and programs go into action and/or existing programs/projects are modified. However, even in the time that it took for a strategy to be developed and action implemented, conditions may have changed. Audit, review, revision, and adjustment are in order. Strategy development and its implementation and review are in a never-ending cycle.

Stategy at the Corporate and Business Level:

In developing the corporate strategy, a matrix has been used to classify various businesses (business units and product lines) of the corporation with regard to market share and industry growth rate to suggest what to do with them (Figure 6). Accordingly, businesses that fall in the area of high growth industry and are likely to maintain high market share are candidates for further investment. In this case, projects for new developments are likely to get support. On the other hand, business units that become classified as being in a low growth industry with a diminishing market share are likely to get limited corporate funding. Projects in this case will probably get support only if they are process oriented with productivity and cost reduction as their objective.

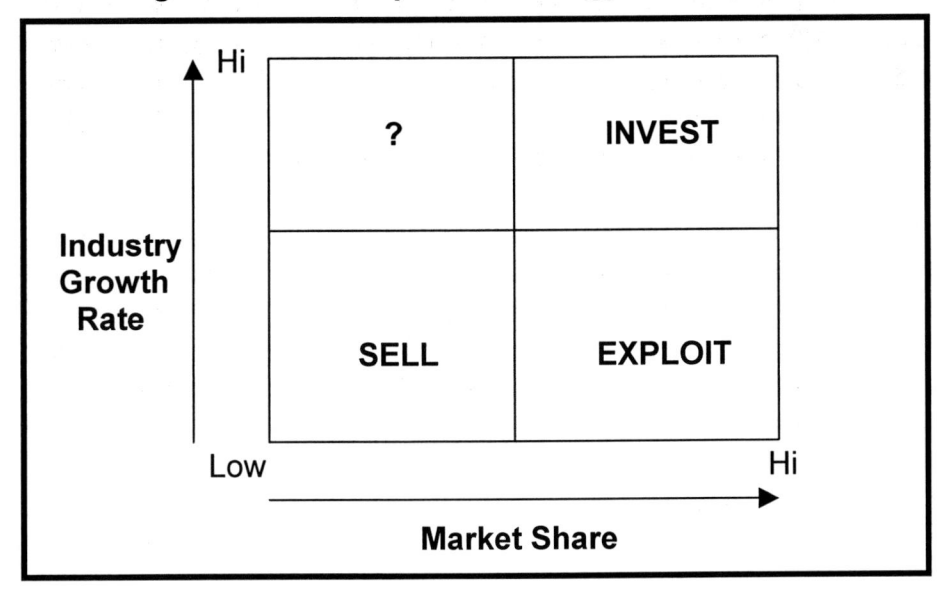

Figure 6: The Corporate Strategy Classification

	Low Market Share	Hi Market Share
Hi Industry Growth Rate	?	INVEST
Low Industry Growth Rate	SELL	EXPLOIT

MANAGING TECHNOLOGY
DEPENDENT OPERATIONS

Another more recent approach is to examine corporate strategy beyond just market share and consider a broader definition of corporate strength. Consider an example of a small market share, but some other aspect such as patents, financial stability, or some core competency that would make the company more secure. Would the development projects have a chance to be supported?

Reflecting back to Figure 2, the business unit's management, having been given directions in terms of mission and vision from the corporate strategy, have a choice of three generic strategies for competing (Table 1). Projects for business units with a generic strategy of differentiation will be supported if they fit a market-oriented objective. It is important in this case, to understand that product differentiation must bring additional revenues. Inevitably, product differentiation will trigger additional development with associated expenses and investment requirement. This form of competition is most often sales driven and subject to market-pull.

Table 1: Generic Business Strategies for Competition

- **by DIFFERENTIATION:**
  Externally oriented efforts...........
  new product or service development at an increasing cost

- **on the basis of LOW COST:**
  Internally oriented efforts............
  process improvement targets resulting in cost reduction

- **through FOCUS:**
  Limiting customer range to differentiate and control costs.

The low cost competitive strategy is likely to be production-process driven. This is typical of businesses in mature or declining industries, which are capital intensive and heavy in technological core competency. This is likely to be a technology-push strategy. It frequently has a simple goal of cost reduction, but is burdened with complex and difficult technical challenges. Nevertheless, it does not have to sound bleak. If well executed, it can change competitive balance within an industry and make the business attractive. A winning strategy may be to pay attention to efficiency while maintaining quality and reputation, thereby to end up as one of limited sources for the product when others abandon the market. Profit margins could recover.

A focused strategy can both differentiate and be low cost, but for a very limited market. It often is the strategy of start-up operations with good new ideas, low overhead costs, and limited access to capital.

In summary, for project managers and project leaders it is important to understand the overall organization's or business unit's strategic planning process at each step of the planning paradigm. The subject of strategic management consideration in a business environment would not be complete without a mention of what has become a management classic. A simple but meaningful scheme to evaluate the strategic profit potential of a business has been proposed by Michael Porter (4). It suggests the following five categories of questions:

> *A winning strategy may be to pay attention to efficiency while maintaining quality and reputation...*

1) What is your bargaining power with the suppliers?
   - Are there only few, each unique and powerful?
   - Are they likely to integrate forward, to take over?
   - Is our particular product of significance to them?

2) What is your bargaining position with your customers?
   - Are you dependent on few or many suppliers?
   - Is their a range of feature and capacity demands?
   - Are they likely to integrate backward, take over?

3) What is the possibility of substitute products?
- Are they likely to be lower price or better performance?
- Are the substitutes likely to come form capital rich source?
- Are there possibilities of complementary products?

4) What is the threat of a new entry – new competitor?
- Are government regulations, distribution channels limited?
- Are capital requirements and other costs a significant barrier?
- How entrenched are existing competitors in the marketplace?

5) Is there intense rivalry within the industry?
- Are there many competitors of similar size?
- Is the battle over market share or market growth?
- Are capacity increases only in big, costly steps?
- Is it difficult to get out of the business?

It should be rather easy to see where a project manager in the role of a leader and an information gatherer can be a significant source of answers to a number of the above questions. If nothing else, a project manager must realize that such questions are being asked. Further more, what about complementary products? Is there an opportunity to develop products that simply are a necessary complement to a product already in the marketplace? Development of new computer chips expands the potential for PC's, while broader use of PC's increases the demand for chips.

Who are the stakeholders in the project?

Today the above questions are as relevant as they were in 1980's. Porter in a *Harvard Business Review* article discusses an expansion of the subject to extend it specifically into the new economy, e-business, dot.com, and information age. It questions the myth that Internet negates the significance of the conventional forces of competition (5).

STRATEGIC ISSUES IN MANAGING TECHNOLOGY
By Donatas Tijunelis

Projects Have a Strategy

Projects can go through a strategy development process similar to what is used in the development of a corporate or business strategy. Consider the following:

A. <u>Stakeholders</u>: Whom does the project serve? Just as in the case of the strategy for a corporation, project strategy development begins with the recognition the stakeholders. While the project may have come from one source, the real stakeholders can be someone else. For example, it may be that another department requested a project to be initiated to serve their self-serving needs and may or may not match the needs of the ultimate user of the product – the business customer. On the other hand, a sales person may request a project to serve his/her particular customer without regard to the stress it places on production or the relative strategic importance of the particular customer to the corporation as a whole.

It is not a bad idea at the outset of a project to list and prioritize the stakeholders in the project and those on whom the project depends. Having recorded the list as part of the project proposal, it becomes useful to periodically review it and more consciously recognize and respond to the changed conditions of the project. List as many supporters of the project as possible.* For example, one may be able to identify the following stakeholders:

- A member of marketing department who identified the initial idea.
- A sales representative whose sales territory is likely to be the first to be tested.
- The engineer in the technical sector who will prepare the prototype.
- Two technicians whose job will be to perform the daily details of the project.
- A university intern who will assist during the summer.
- Production plant manager whose plant may get more orders if the project succeeds.
- The director of advertising.

MANAGING TECHNOLOGY
DEPENDENT OPERATIONS

- The manager of marketing or who formally requested the project.
- The supplier of the key component.
- The custom shop that will fabricate and assemble the required sub-components.
- The lawyer who is to investigate trade name and patent issues.
- The financial manager who allocated funds and prepared a budget for the project.
- The outside consultant or reference, which confirms the technical feasibility.
    * These can even be prioritized by a pair-wise ranking technique given in the Appendix

With a listing of stakeholders in hand one can develop a project "selling" plan as well as use it for subsequent project control through periodic inquiries for feedback.

B. Mission: What is the mission of the project? Having considered the issue of project stakeholders, it is necessary to quickly and briefly recognize the strategic mission of the project. How does the project serve the mission of the organization and the corporation. Is it in line with how the corporation wants to be identified? For a company that values reputation, the project better be consistent with good business practices. Its mission must be to contribute to that reputation.

C. Vision: What about the project's vision? Here the task of establishing a vision for the project is more challenging. In this case, the project has a dual role – reactive and proactive. In the reactive role the goals are simply handed to the project manager. If they are succinct and specific they must of course be followed to formulate project plans. However, at the same time, they have to be understood in the context of a broad vision statement and in line with the functional business goals and corporate vision of its future. Check to see if the vision of the project results will be interpreted in the same way as implied by demands of the broader strategic plans.

D. SWOT Analysis: Are projects internally and externally justified with respect to a broad business strategy? The SWOT approach to examine internal and external factors can be adapted to project evaluation. Internal questions are those regarding competency, physical resources, infrastructure, time allowed etc. External issues to a project may be those of access to market data, upper management's indecision, an alternative plan, alternative new technology etc. The business condition strengths and weaknesses and perceived external factors lead project management to ask the questions of the fit of their projects as shown in Table 2.

> *...project leadership is to have a unique opportunity to recognize project's potential and risks*

Table 2: Project Questions At Varying Business Conditions

POSSIBLE BUSINESS CONDITIONS		STRATEGIC QUESTIONS TO CONSIDER:
INTERNAL	**EXTERNAL**	
• STRONG COMPETENCY @ an ADVANTAGE	• CLEAR BUSINESS OPPORTUNITIES	❑ Do projects implement an aggressive growth strategy?
• WEAK COMPETENCY @ a DISSADVANTAGE	• CLEAR BUSINESS OPPORTUNITIES	❑ Do projects support turnaround strategy?
• STRONG COMPETENCY @ an ADVANTAGE	• CLEAR BUSINESS THREATS	❑ Do projects implement diversification strategy?
• WEAK COMPETENCY @ a DISSADVANTAGE	• CLEAR BUSINESS THREATS	❑ Do projects support a defensive strategy?

Project justification is easier with a clear understanding of project fit with the corporate, business, and functional strategy. To answer the above questions, a survey of the key project stakeholders may help by asking the for the following inputs:

1. A list of existing capabilities within the organization or business that strengthens a project in support of a strategy. Give specifics of how to take advantage.

MANAGING TECHNOLOGY
DEPENDENT OPERATIONS

2. A list of organization or business weaknesses that may create risk of project failure. Show how to overcome.
3. A list of new external opportunities that the project may bring up. Suggest avenues for exploiting.
4. A list of threats to the project from outside of its scope. Are there new alternatives for the support of business strategy? Indicate how can the potential impact be minimized.

Armed with a list of identified corporate SWOT factors and having answered related questions with respect to the project, the manager of the project in a leadership role is better prepared to justify and/or "sell" the project.

E. Project Support, Implementation: How to justify or "sell" a project? Here the issues are of specific project value to a business need and how it is presented for approval and support. Project selection and project evaluation go hand in hand since there may be a range of project options for a limited budget.

When projects start from within, the principal issue is that of channeling and exploiting creative thinking toward a yet unsubstantiated value of its result. In another case, the desirability of the result is known but the challenge is in the choice of several project options to achieve it. The project can also be purely tactical with a prime challenge to simply marshal all of the resources available to get the job done. Finally, there may be projects to build core competency in an area of strategic importance. These are the product of organizational development and long range strategic planning. Therefore, projects can be classified into four main categories: projects that start with an idea from within; projects which are assigned from outside; crisis projects with a survival goal; and projects to build skills:

    1. Projects from within can start as-
- new exploratory idea outside current interest
- consequence of on-the-job experience
- projects which arise from ongoing development programs

2. Projects from outside respond to-
· new demand by a customer
· new product entrant in the market
· change in management strategy
3. Crisis projects-
· unexpected internal event giving no options
· driven by government regulations, law suit, etc.
4. Skill building projects-
· preparation to expand into new field
· sustaining valuable core competency

Projects that start from within have to be marketed. They have to be sold to gain support, to be given proper resources, time, and scope. Their results have to be defined in terms that will attract initial attention and serious review. This is not easy to do. A project that comes from within has cost as the most obvious feature that has to be sold to management. After all it is not their idea. The benefits have to be described in the terms that upper management's decision-makers can easily appreciate.

Projects that originate from an outside assignment may appear to be specific, but often it is the result of a broad need that the management of the organization has recognized. It may be part of a strategy based on a planned program to be implemented through a cluster of projects and a choice of several project options. The benefit of the overall effort is recognized and accepted by the decision-making management, but the implementation cost, time, and scope are yet to be made evident and "sold." Strategic fit of an individual project or some of its features may be undetermined and not clear to all concerned.

> *Strategic fit of individual projects may be undetermined.*

Crisis projects are unique in that by definition time is of the essence, cost becomes secondary, and scope can become more flexible. The challenge is not so much in planning, as it is in efficient deployment of resources, execution of tasks, and reporting progress. The cost / benefit relationship of the project in this case is almost totally the responsibility of top management. Crisis projects can arise internally, as in the case of fire in a plant, or from outside, in the case of computer virus, or as a result of the emergence of a disruptive technology. The latter merits an explanation.

Most of the internal and external projects are based on sustained technology extensions. They may be new and novel, but are to some extent built upon prior knowledge. However, once in a while there appear disruptive technologies that are pure discoveries, totally unknown, and therefore unpredictable in any form. The computer disc drive industry is an example of disruptive technology that made planning very difficult. Under such circumstances it may be necessary to simply rely on an educated staff that is versatile and will be able to respond when faced with the emergence of a disruptive technology.

> *Some projects need "selling"...*

Skill building projects may require "selling" on the part of project as well as general management. In the fast pace of technological changes, constant learning is essential. Some projects can be seen as essential to learn and experiment with the extension of existing techniques. This need may be obvious at the project level, but difficult to recognize by general management. It needs "selling" upwards. In practice, project managers do not like the "selling" part of their job. It is, at times, taken by them as demeaning. Here project managers must step into their leadership role.

On the other hand, general management may see a need for a major shift in an organization's vision of the future. Their longer-range strategic thinking may not be evident at the functional level of the organization. A program of projects to build new core competency in a promising field not yet clearly identified needs "selling" to the technical staff. Focused on their present tasks, they may become confused, see their work as useless, and lose motivation. Some projects may be requested simply to sustain core competency in an area that is important but not intended to grow. It may be misinterpreted as a loser. Here too, some "selling" may be necessary since project benefits can be seen as career limiting. The project leader must sense the possibility of the problem and ask for more direct communication from the top.

Project Evaluation

Evaluation of a project is a part of its justification. For justification and selection purpose both quantitative and qualitative data is used.

Using Purely Quantitative, Capital Budgeting Techniques: Planned projects are budgeted. Industrial budgeting practice uses capital budgeting techniques. Therefore, some value measure is expected for a contemplated project in similar terms. In practice, investment value is judged by return on utilized assets. Measures such as internal rate of return (IRR), net present value (NPV) and the pay-back period are most popular. The financial accounting principles which calculate these are readily available and the actual data manipulation can be done using built-in features of common computer spreadsheet software.

> *... IRR and NPV are inherently biased against long-term ...*

These standard techniques are tempting. They sound definitive. Pressured by time, managers with an accounting or financial backgrounds find comfort in financially based crisp decisions. Such project value measures for them are easy to understand and justify. Yet, calculations such as IRR and NPV can be inherently biased against long-term investments in new technology. The return hurdle rates (interest) mirror current venture capital and bank rates. They are set, published, and obvious. On the other hand, returns based on the success of a project are subjective and unsubstantiated.

MANAGING TECHNOLOGY DEPENDENT OPERATIONS

They depend on many factors that have an impact during the project's life-cycle and only a few people are available to explain them. Commitment seems riskier. If nothing else, projections of costs and returns are probabilistic and management's willingness to mitigate risk is hard to come by.

A precise financial evaluation based on limited, purely financial projections of returns and costs can lead to a false sense of accuracy. The decision-makers considering projects purely in financial terms may quickly dismiss or with false confidence accept the "sale" of a project presented in these terms. That being said, some measure of return for an investment is necessary. To have a measure of financial impact, such as break-even analysis and pay-back time, is obviously important. However it should not be the sole determinant of a strategically targeted project's value.

> *...at a minimum, some form of pay-back analysis should be carried out.*

Given the monetary value of an expected strategic outcome or the Expected Monetary Value (EMV), a decision tree technique can be adapted for a quantitative project ranking. With more than one project option to achieve a strategic objective, the choice for highest merit, lowest cost, and minimum risk can be identified. This is done by calculating backwards through the probabilities and likely options at each stage of each project's life cycle. Assignments of probabilities are made for each stage. It then becomes simply a mechanical technique to calculate the cumulative monetary value at each stage of the innovation process to make the final ranking of options. Many project management software programs, as well as operations management texts, have accompanying software that perform decision tree calculations automatically.

> *Business-simulation tools can help...*

For those that favor financial analysis in the form of capital budgeting tools, some business simulation software may be helpful. Business simulation tools can help map out decision effects. Project managers perhaps can better see what impact a major project or a portfolio of projects can have under a range of simulated financial and market alternatives. The key to getting the most from a simulation is to know what your objective is and recognize the built-in assumptions. Harvard Management Update offers a sampling of available tools (computer software), with advice on their uses and limitations (6).

Evaluation Probabilistic Models and Scoring Techniques: As alternative to the purely financial measures, quantitative approaches specific to project evaluation seem to cycle in and out of favor by technology management advocates. The industrial Research Institute in Washington, DC publishes *Research-Technology Management* that frequently features articles dealing with project evaluation metrics. Some can have direct application if the model features match your case. However, many are more useful as an audit of technical programs to help future management planning rather than as a tool for on-the-spot decisions.

> *...here the feature of probabilities and subjective factors are included.*

Literally there are hundreds of models for R&D project evaluation, most of them have never been adopted for routine use (7). In the 1970-80's there was extensive government support for development programs, vigorous growth of operations research as a discipline and, as a result, a trend of self-justification for extensive strategic planning. Enthusiasm was further fueled by the anticipation of available computer power and access to large data banks. Complex mathematical models were suggested specifically to evaluate projects.

However, "garbage in leads to garbage out." Availability of meaningless data collected without full consideration of a variety of indeterminate subjective conditions and often as only extrapolation of the past led to a false assumption of accuracy. Furthermore, to refine the data and carry out comprehensive reviews of complex mathematical solutions seeking one best answer became time consuming. The planning and evaluation process fell out of step with the pace of the changing technological and economic environment. Top management decision-makers could not spend the needed time to understand all of the parameters of the evaluation models. It led to indecision or to after-the-fact theoretical exercises.

What may be important to one business may not be important to another. A manufacturer will place a lot of value in capacity utilization while a service organization may pay a lot of value on response time. Do not use a quantitative model of project evaluation of the shelf and expect it to fit exactly. To be practical, look to a simpler and more meaningful approach that you can adopt for your circumstances.

> *...scoring models are custom designed.*

Consider project scoring. Scoring models consist of a number of criteria with a scale customized for each business to help evaluate and compare projects. They can be a combination of intuitive judgments and a structure to include some quantitative facts that are relevant to the business. The key is to create a scoring system that is weighted according to what is important internally to the business owners, the prime stakeholders. It should have buy-in, take into consideration probability estimates by those most knowledgeable, and be part of a record of commitment and prioritization until the next review.

For a small business unit, it can be made very simple and be used primarily in a project review process to see how project value may have changed.

STRATEGIC ISSUES IN MANAGING TECHNOLOGY
By Donatas Tijunelis

Scoring models are useful for a diversified corporation with more than 25 projects whose technical commitments to operations are monitored centrally. A subjective score facilitates the prioratization of a large project list that is part of one business entity budget. As such, it can provide help to screen out, on a common basis, projects for more in-depth review and decisions, as well as be a tool in the process of continuous project portfolio improvement. The weighting factors and the resulting score are arbitrary and have little absolute value. They are only to focus decision-making within the business "family" on what ranks high and what ranks low.

One example of a scoring technique is the Probable Economic Index (PEI) used by a large manufacturer to review and rank 45 technical projects. In this particular case, a planning committee was composed of representatives of various stakeholders. There was a member of production management, a member of technical development, a representative of management from three different business units and corporate marketing. Their objective was to establish scoring criteria that meets common corporate goals. Having to consider the support of a number of projects for both process and product innovation, they established a scoring system as follows:

$$\text{PEI} = P_t (V_a + V_b + V_c + V_d) + X$$

Where:

P_t = Total Project Probability = $(P_f)(P_m)(P_s)$
  P_f = technical concept feasibility, (0 to 1.0)
  P_m = manufacturing, (0 to 1.0)
  P_s = sales, (0 to 1.0)
V_a = Target Market Share Gain:
  1 for each $100k gain/year
V_b = Target Gross Margin or Savings in 3yrs
  3 if >40%; 2 if =20-40%; 1 if = 10-20%; 0.5 if <10%
V_c = Target Development Costs
  3 if <$150k; 2 if =$150k-500k; 1 if >$500k

V_d = Capital Requirements for Production
3 if <$25k; 2 if =$25k-75k; 1 if =$75k-150k; 0.5 if >150k
X = Commitment to a Customer or Negative Proprietorship Issues 1 to –3 respectively

    The probability of meeting the intended objectives (P_t) was to be estimated by each project's manager through independent inquiries of the management most responsible for the implementation of manufacturing, sales, and of the technical staff of the project. Within a month, PEI for all projects was calculated, the contributing factors described and submitted for review. In reviewing project portfolio through this simple approach, the company found the following:

1. A relatively attractive, highly-promoted project aimed at large market share with high sales probability was found to have less than 10% chance of success. Its P_s of 0.8, when multiplied by independent estimates of probabilities for technical feasibility with P_f = 0.2 and by scale-up to manufacturing with P_m = 0.5, gave a P_t of only 0.08.
2. There was an unexpected relatively large difference between the PEI scoring of the projects in their total program. They ranged from a promising 5 down to less than 1 for the very risky ones, allowing more objective attention to the extremes.
3. The process itself, because of its simplicity and interdisciplinary involvement, generated awareness of the issues strategic to each project and allowed for an evaluation of the development program as a whole.
4. It became easier to get top management's attention. Follow-up financial analysis became better focused.
5. The "selling" of good projects became easier and the abandoning of poor ones became better appreciated by project participants.

> *Precise results provided too late are of little value.*

    With regards to precision, it should be obvious that the factors of probable economic impact scoring and evaluation can be made more precise by examining subsets. For example, the probability of manufacturing, Pm, could have been the product of (Pm-1)(Pm-2)(Pm-3) representing process step estimated by the raw materials handling department, assembly supervisor and inventory manager, respectively.

Given the opportunity for more detail in the input data, the results can be made more precise, but in some cases that is not necessarily what is needed. It takes more time and can stall decision making, which can create an anti-entrepreneural climate that is negative to project implementation. Precise results provided too late are of little value. They can introduce unexpected sources of bias and arguments and in the end can be less useful – not worth the effort. It is up to the project manager to judge what is reasonable under the circumstances.

Intuition, Nominal Group Techniques, and Subjective Prioritization: Various forms of brainstorming can be used to develop a form of ranking. In general, group based decision is better than a single opinion or a numerical projection based on inappropriate data. However, there are two tactical problems. First is that it is difficult to manage a brainstorming session to compare more than 25 diverse projects. Personal interests and company politics can play havoc. Second is that if intuition if not assisted by some background information and structure for comparisons it can be very arbitrary. Of course, whoever provides the background and structure can introduce bias.

> *...brainstorming can be used to develop a form of ranking*

For some structure without bias, ask the participants to examine each project through a SWOT analysis before forming their opinions.

For an objective background, examine the issues of the project's environment in the context of the competitive forces facing the business organization as a whole. For example, for project evaluation the review of competitive forces may cover the following issues:

1. **Demand conditions and power of the client**
   - Will the client define the goals correctly; will they change?
   - Will you be informed to adjust?
   - Are there other clients waiting for the same results?
2. **Resource reliability**
   - Is your resource unique and stable?
   - Can the project be subdivided and parts outsourced?
   - How can clients limit your options?
   - Are there joint venture opportunities?
3. **Stability of business climate**
   - Are you informed of the business conditions?
   - Is the current funding of the project reliable?
   - Can the direction and outcome of the project affect business climate, at what point in the innovation process?
4. **Possibility of another option, approach**
   - Are there alternatives to your project available?
   - Can the business afford them?
5. **Risk of internal social issues**
   - Does the project depend on government regulation, patent, contract issues yet to be clarified?
   - Are there internal labor and skill commitments in question?
   - Are there personality conflicts on the project team?
   - Have territorial claims within the organization been resolved?

When there are less than 25 projects to evaluate, or for a smaller organization with few projects for very limited staff and budget, a brain storming session may be most reasonable. It should be composed of well chosen stakeholders, as discussed earlier in the chapter, and rated by a pair-wise ranking technique. An example with details of this simple technique is given in the Appendix. It is designed to compare in pairs all ranking options. Its value is in recognizing that people are likely to be more objective in their intuition of what is more valuable when they make one-to-one comparisons rather than comparing all options at the same time. This is especially true when projects have diverse features to be evaluated. The use of this tool reflects intuitive knowledge and is structured enough to come to a quick ranking of a group of projects. Proper selection of the participants, the stakeholders, is very important.

Beyond the simple, nominal-group brainstorming approach, a subjective, yet more structured and very rational method of analyzing and simplifying decisions of relative merit is the Even Swap method (8). This technique can be used to compare project's merits for various business objectives. It encourages speculative comparisons of values on a relative basis between dissimilar features in order to eliminate lower priority options. Regardless of its intent to be simple and logical, the techniques can become complex and could be difficult to present. For a small operation with few project options, a simpler, yet disciplined, approach may be advisable. In such cases, scoring models are custom designed. One of these is described next.

> ...*no single prescription of how to evaluate and select projects.*

There is no one best way to manage and lead a project. There is no single prescription of how to evaluate and select projects. It all depends on the business, the organization, the economic climate, and the social considerations of those involved. Each program and each project should justify its own evaluation scheme. It is up to the management and its leaders to identify what is important to them in view of the strategy and the circumstances at hand. The majority of management is forced to make basically intuitive decisions with little or no systematic analysis. First-mover's advantage became the result of leadership with a vision of future scenarios and not mathematical extrapolation of past data. In the age of discontinuities, practical intuition that takes in subjective factors became more appropriate. Jack Welch's very successful leadership of General Electric Corporation without the adherence to the mechanics of strategic planning, relied on strategic thinking and became exemplary. For practical purposes quantitative

MANAGING TECHNOLOGY DEPENDENT OPERATIONS

> *...forced to make intuitive decisions.*

evaluation of projects has been sidelined. Now, a new approach is emerging stimulated by a simple yet effective, look at the circumstances and at an organized vision of possibilities.

Evaluating project options for strategic impact prioritization is the key. In the area of manufacturing, Eliyahu M. Goldratt brought to the attention the benefit of simplification (9). Project managers should focus on a few critical areas, not divide their attention over all of the tasks. In a 1998 review of project management, Elton and Justin Roe suggest to "examine some important drivers of project performance: namely, how well new initiatives align with strategy and how successfully an organization balances its overall capacity and capabilities with its portfolio of projects." In their way of thinking, an essential element fostering collaboration are project managers who can handle political as well as the technical aspects of their projects. Leadership built into a project plan may become the main constraint. "Projects incorporating discontinuous technologies by definition involve work outside companies experience. In such case, a detailed project plan would give managers a false sense of security, but if they pay attention to the constraints, they will be on their way to capturing and consistently managing the risks of their projects" (10).

Evaluating Projects for Global Marketing and Support: Global business ventures, such as exports or the search for foreign direct investors, will have some additional aspects of importance to deal with communications, government regulations, appreciation of cultural differences, etc.

> *The most obvious requirement of a project leader is to be an entrepreneur*

There is one notable additional difference to be recognized by the project manager. At the business management's commitment level, there are many more questions to be answered for the potential stakeholders. That which is obvious domestically with regard to the worth of a project may not be clear to a potential foreign importer or investor. When trust is weaker, more information and factual data support is needed to "sell" and manage a project.

International project management is a subject that should be treated with special attention. As a start, a textbook such as *International Management* by Charles Hill is worth examining for general appreciation of the issues involved (11).

The Strategic Thinking Role of a Project Leader

When major change and innovation is the operations intent, the project leader, should be proactive to generate scenarios of potential opportunities and to synthesize ways to capture them. In the context of Total Quality, the greatest value may not be in contributing to the details of an operation but to the input to the strategic thinking process. The difference is well defined by Henry Mintzberg (12):

"Strategic management isn't strategic thinking. One is analysis, the other is synthesis"

The issue is whether the problem is to be assigned, offered to the project manager, or sought after as an opportunity for the future by the project leader. Solutions to a series of problems show a pattern. Problems need to be anticipated. In the fast pace of technological, political, economic, and environmental changes, successful entrepreneurs strive to identify problems and opportunities in advance. A project manager should seek to expand into the role of project leadership as a building block for change.

The most obvious requirement of a project leader is to be an entrepreneur – to be flexible and adaptive. "Management is about coping with complexity. Its practices and procedures are, for the most part, responses to the emergence of large, complex organizations in the twentieth century. Leadership, by contrast, is about coping with change. Part of the reason

MANAGING TECHNOLOGY DEPENDENT OPERATIONS

leadership has become so important in recent years is that the business world has become more competitive and more volatile. More change always demands more leadership"(13). This requires the project leader first to know and understand corporate strategy, concur with its vision, and then to be effectively resourceful to respond – taking responsibility for project justification, anticipate its implementation throughout the innovation process and make hard decisions, as needed of a leader.

Studies of successful projects show that powerful guiding visions of business, project, and product strategies were mutually reinforcing, energizing the people on the teams, focusing attention and effort on the right things, and getting them done in the right way. Middle managers, controlling the destiny of several projects, invariably are put in a leadership role. They need to evaluate projects relative to each other, prioritize them and adjust according to changes in business strategy. It is a tough job.

> *...managers take projects well past the point at which they should drop them*

Some managers take projects well past the point at which they should drop them. "New initiatives often gain momentum even as it becomes clear that they are doomed. The reason: blind faith in their success." The quote is from a well-titled article "Why Bad Projects Are So Hard to Kill" by Isabelle Royer (14). Project leaders must look closely at themselves and recognize what influences them. Some influences are psychological—they've been rewarded in the past for sticking to their guns, so why shouldn't the same thing happen this time? Some are social —no one likes a loser. Some are structural i.e., important members of the organization are publicly committed to the project.

It is important for a project leader to recognize management's span of attention as illustrated in Figure 7.

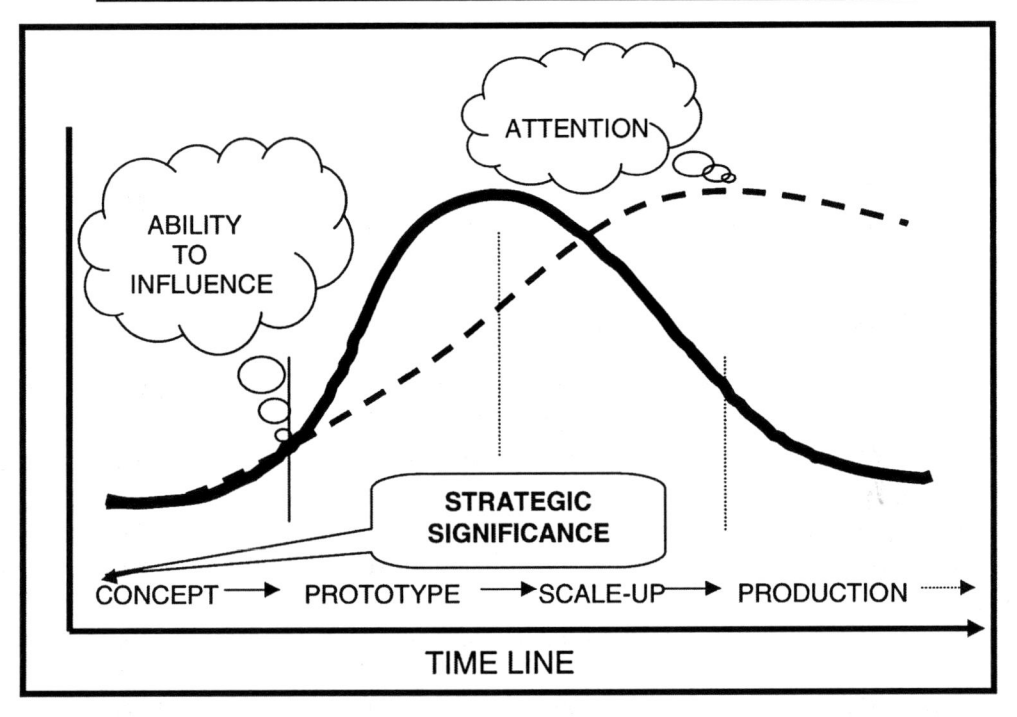

STRATEGIC ISSUES IN MANAGING TECHNOLOGY
By Donatas Tijunelis

Figure 7: Leverage and Attention in Project's Life-Cycle

The need for attention may be obvious to some who manage projects at hand, but not obvious to others. There is an apparent paradox between attention and consequences that has to be recognized. When costs are low and change is easy, attention at the top is lacking. When costs become high, upper management notices, but changing the course is costly – projects tend to go on in spite of less than optimal consequences.

Earlier in this chapter it was noted that operations management has two features. One is that of effectively maintaining steady state and the other is the ability to manage change in line with market needs and availability of resources. Steady-state management requires a tactical approach. In this case, project management accepts the objectives of the stakeholders and organizes the resources. That means to insure that the quantitative aspects of resource planning, scheduling, deployment, and cost controls are implemented effectively. Commitments need to be kept. A project manager

MANAGING TECHNOLOGY DEPENDENT OPERATIONS

will be asked to insure that the rigid dependence between time, scope, minimum cost, and the resulting projected performance will be maintained.

It is the responsibility of a project manager to recognize that if one of these factors changes something else will have to give. The impact of the changes can be overlooked and may be strategically significant. It should be the task of the project manager in a leadership role to be prepared to point out the strategic long-range impact of the problem.

> *The manager does things right, the leader does the right things!*

In conclusion, strategy is either top-down planned or emergent or, to be real, the combination of the two. Planned strategy is formal and guided by general business management – the office of the president, the CEO, or in other cases the owner, who takes the role of general manager. The emergent strategy bubbles up from the "troops." It may come from input of business managers and the market place through the sales organization. It can come through the technical management as tips of technical innovation opportunity that a technician has found at a technical exhibit. It could also come from a machine operator in a plant who points out possible productivity improvements to the plant or operations management. Synergistic interplay of diverse inputs is what makes an effective technology dependent operation. Its up to the leadership of the organization to recognize the value of different points of view and to establish a process that makes the idea exchange work to answer the following questions:

   I.   Where do we stand now?
   II. Where do we want to be?
   III. How do we get there?

In considering project evaluation and selection for a fit with an organization's strategy the following should remind project managers of the need to recognize their broader role:

The manager does the things right, while the leader does the right things.

References

1. Featured by Ted Koppel, "IDEO – The Deep Dive", *ABC Nightline, 1999*, www.ideo.com

2. Stefan Thomke, "Enlightened Experimentation: The New Imperative for Innovation", *Harvard Business Review*, Harvard Business School Publ., Boston, February, 2001

3. Charles W. Hill and Gareth R. Jones, *Strategic Management: An Integrated Approach*, 6th ed., Houghton Mifflin Co., Boston, 2004.

4. Michael Porter, *Competitive Strategy*, Free Press, Mew York, 1980.

5. Michael Porter, "Strategy and the Internet", *Harvard Business Review*, Harvard Business School Publ., Boston, March, 2001.

6. Harvard Management Update, "Should You Be Using Simulations?", *HBR, Harvard Business Review*, Harvard Business School Publ., Boston, 2000.

7. Selected Papers, "Selection, Control, Evaluation and Termination of Projects", *Research Management*, Industrial Research Institute, New York, 1973-1980.

8. John Hammond, Ralph Keeney and Howard Raiffa, "Even Swaps: A Rational Method for Managing Trade-Offs", *Harvard Business Review*, Harvard Business School Publ., Boston, March-April, 1998.

9. Eliyahu M. Goldratt, *The Goal.* . 2nd ed, North River Press, 1992.

10. Jeffrey Elton and Justin Roe, " Bringing Discipline to Project Management", *Harvard Business Review*, Harvard Business School Publ., Boston, March-April, 1998.

11. C. Hill, *International Management*, 5th Ed., McGraw-Hill, NY., 2003.

12. H. Mintzberg, The Fall and Rise of Strategic Planning, *Harvard Business Review*, Harvard Business School Publ., Boston, January-February, 1994.

13. John P. Kotter," What Leaders Really Do", *Harvard Business Review*, Harvard Business School Publ., Boston, 1990.

14. Isabelle Royer, "Why Bad Projects Are So Hard To Kill", *Harvard Business Review*, Harvard Business School, February, 2003

Suggested Additional Reading

- C.M. Christensen, *The Innovators Dilemma*, Harper Collins Publising, 2000.
- Selected Papers, "Selection, Control, Evaluation and Termination of Projects", *Research Management*, Industrial Research Institute, New York, 1973-1980.
- Jeffrey Elton and Justin Roe, " Bringing Discipline to Project Management", *Harvard Business Review*, Harvard Business School Publ., Boston, March-April, 1998.
- Hodder, James E.; Riggs, Henry E., "Pitfalls in Evaluating Risky Projects", *Harvard Business Review*, Harvard Business School Publ., Boston,, 2001.
- J.W.Higgins, *101 Creative Problem Solving Techniques*, New Management Publ., New York, 1994.
- Patrick J. Bellow, George L. Morrisey, Betty L. Acomb, *The Executive Guide to Strategic Planning,* Jossey-Bass Publ., San Francisco, 1989.

STRATEGIC ISSUES IN MANAGING TECHNOLOGY
By Donatas Tijunelis

APPENDIX

Pair-Wise Prioritization Technique *

When choice has to be made for the relative significance of strategic issue factors among many, each of which having many subjective features, it is difficult to do. Comparisons of a feature of significance to one issue will have different value to another. Each issue has a relative overall value whose comparison can only be done in the time allowed reasonably well to one other, but not to many at the same time. For example, consider comparing a project for the purpose of maintaining skills vs a project for a specific customer to gain an immediate sales order. Further complication may develop if at the same time there is a rush for another new project that requires attention from the same person who already feels overworked. What if there is another project that has great promise but its cost could be prohibitive, and one that the company president wants, and several other projects with various subjective features. What to do? Consider the following example; it works easier in practice than it looks:

Example

1. Examine the list of issues to be prioritized.
2. List the same for the vertical row and the horizontal column headings.
3. Compare issue #1 against issue #2 in a one-one comparison.
4. If #1 is more significant than #2 place an X under #2 in row 1
5. Repeat comparing #1 against #3. If less significant place 0 in the columns column 3, row 1.
6. Continue till all blank spaces are filled.
7. Add all the X's horizontally for each row, all the 0's for each column. List at bottom.
8. Add the X's to the 0's for each column.
9. The largest total represents the highest significance.
10. Repeat the process for the top ties. If there are many ties, eliminate lower ranking ones or do a more detail analysis of factors of those equally ranked

MANAGING TECHNOLOGY
DEPENDENT OPERATIONS

Pair-wise Prioritization Example

	1	2	3	4	5	6	7	8	9	X
1		x	0	x	x	0	x	0	0	4
2			0	x	0	0	0	0	0	1
3				0	x	x	x	0	x	4
4					0	0	x	x	0	2
5						0	0	0	x	1
6							x	0	0	1
7								x	x	2
8									0	0
9										0
0	0	0	2	1	2	4	2	5	5	
X	4	1	4	2	1	1	2	0	0	
Σ	4	1	6	3	3	5	4	5	5	
RANK			#1			#2			#2	

* Based on *Strategic Areas Desision Matrix* by Patrick J. Bellow, George L. Morrisey, Betty L. Acomb, <u>The Executive Guide to Strategic Planning</u>, Jossey-Bass Publ., San Francisco, 1989. Original source: Copyright © George Morrisey, 1974.

Chapter 10, Lead-in:

TECHNICAL SUPPORT TO MULTIPLE OPERATIONS

In this chapter the book takes two tracks:

First, it gives the non-technical manager a view of how technical managers prepare to provide technical services to the business units. It should give some insights on the tips technical staff share among themselves to make their function more effective. It relates the experience of the technical managers and the technical staff of a successful, large, diversified corporation.

Second, the chapter provides as an example of what to consider when managing a function whose primary concern is technical support of a large variety of operational situations. A "checklist" of 50 tips/tools to consider when examining technical support commitment to range of operating business units is included.

The example is based on the unique practice at Illinois Tool Works, Inc. (ITW) Technology Center. ITW is a multi-billion corporation of some 600 business units worldwide, made up of a variety of sizes and products. Therefore the choice of ITW covers in one example, as much as possible, the technical support issues for both large and small businesses. For instance, some of the tips/tools given in the "checklist" can be useful to the those planning to support a small entrepreneur without production experience or a production plant without analytical capability to respond to an environmental regulation challenge. It shows how to blend practical interests of business operations and those of technology functions. This chapter can be considered in general terms as a model to be modified for other situations.

CHAPTER 10

TECHNICAL SUPPORT TO MULTIPLE OPERATIONS

James P. Nelson

Most of the mid-to-large size companies with multiple product lines have divisional organization structure. The business units and operations are distributed throughout many locations. Some of these may be large and relatively autonomous. A business unit may contain within it many of the core functions such as production operations, human resources, marketing and sales organizations for its line of particular products or the region it serves. On the other hand, it is likely to be tied to the home office or corporate headquarters for strategic targets, legal issues, and financial control.

> *Technical support of day-to-day operations is usually located locally... but may have to extend beyond the immediate ... engage in product or process developments...*

Technical support of day-to-day operations is usually located locally at plants, service centers, distribution hubs, or warehouses. Some of these technical support functions may have to extend beyond the immediate needs and engage in product or process developments leveraging the business unit's competitive advantage in the markets it is intended to serve according to its mission. Very large business units, such as those of GE Plastics or the

TECHNICAL SUPPORT TO MULTIPLE OPERATIONS
By James P. Nelson

Cargill-Dow joint venture, beyond day-to-day support each has an extensive R&D and Enginering of their own. Corporate headquarters of large, multiplant corporations will have a Corporate Technical Center with the primary goal to focus on long-range strategic objectives. They may also provide technical support to operating units as neded.

Some highly diversified corporations, while large in total, are composed of both large and small business units. The size and technical functions of these business units vary greatly. Some units may resemble an entrepreneurial start-up with creative technical staff, but without high volume production experience. Others may be the reverse - primarily a production facility or a distribution center with little advanced product or process innovation capability. Some of these isolated units may also lack the capability to respond adequately to emergency situations, product quality problems, equipment upgrades, and safety or environmental regulation issues. They may resemble a small single-product family-owned business. Under such situations the corporation maintains a central technical resource for both the long-range and short-range objectives. The management of these two objectives is different. The skills and interests of the employees within the Technical Center are different. The long-range technical activity is left to the scientists or scientifically inclined engineers, while the development and technical support is left to applications oriented engineers and technicians. It is the latter that is of most concern to operations management. Companies like ITW Corporation, Borg Warner, and Dow have developed management systems peculiar to their needs. Some companies charge business units a "tax" that covers both technical support as well as R&D activities. Other companies have the business units pay for corporate services when utilized, i.e. project by project.

The experience of Illinois Tool Works, Inc. (ITW) is worth noting. It is a big corporation with many small business units. ITW has been using successfully a system of prioritization for supporting projects – a challenge that all face when dealing with technology dependent

> *...a big corporation and small business units - all in one...*

MANAGING TECHNOLOGY DEPENDENT OPERATIONS

operations. The ITW Technology Center provides technology solutions to the many independent business units of ITW. The Tech Center functions as "Technology-On-Call" for the business units. The Tech Center is funded by a small corporate tax on each business unit, so the business units pay for Tech Center services whether they use them or not. Thus the business units do not pay directly for Tech Center Services on an hourly or project bases. They may be asked to share project costs above a certain level, or for outside services above some moderate level. This serves not only to keep the Tech Center budget whole (How could they budget for all the consultants that 600 businesses might hire?) and as a reality check whether the business unit is really interested in the project. It may be too easy to spend OPM (Other People's Money) and pursue unworthy projects.

One of the tools that ITW employs is a simple checklist to monitor projects. This checklist was constructed to guide engineers in the ITW Tech Center in their role as project managers. It was compiled from a list of "best practices" submitted by almost all of the engineers in the Materials Research Department of the Tech Center. It was originally designed as a simple checklist. The last section of this chapter presents the "List" in a format that can be used as a checklist by the reader.

> *The Tech Center considers the Business Units to be their "customer".*

At ITW, business units have complete freedom to get technology they need however they wish, including using consultants, universities, or engineering firms. They are not required to use the Tech Center. The Tech Center considers the business units to be their "customers". The engineer reading this is strongly advised to identify their key customer. This may well be the actual customer for the new product, but it is more likely, especially in larger companies, that the primary "customer" for R&D services is marketing, business management, or manufacturing inside the company. This customer orientation brings the researcher into a mindset that is congruent with the user of the product of the research – information. This

TECHNICAL SUPPORT TO MULTIPLE OPERATIONS
By James P. Nelson

helps keep the project on a fast timeline. In turn, this helps keep things simple, fast and cheap. In the long run, R&D funding in any company is a function of value delivered. Your customers in Marketing and Manufacturing will talk to senior management about their experiences with R&D. A valued "customer" is likely to get R&D services, which have real value, and give the positive feedback to senior management, leading to continued or even enhanced funding. Identify your customer!

The ITW corporate culture has low levels of bureaucracy and few written reports. Engineers operate independently with little supervision. They work increasingly as teams if projects require a wide range of skills but often work alone with the business unit staff. We expect readers will have a wide range of working environments. The readers should put the comments here into the context of their individual environment.

BEFORE STARTING PROJECTS

Carefully choose and organize: The beginning of the project is the most critical period, but it is often handled too casually. This is the point in time when project direction can go wrong in a myriad of ways. Some of these are discussed below with suggestions for resolution.

1. <u>If a business unit is new to you or the Tech Center, be sure to visit it early, and often,</u> to build relationships and an understanding of what the Tech Center has to offer. The business unit will be more open to Tech Center help if personal relationships are good. If you gain the respect of the business, they will entrust you with future critical programs. A good way to start is with a few fast, simple "Tech Service" jobs. This lets the business unit learn about what the Tech Center can do, how it works, and builds individual relationships.

> *The beginning of the project is the most critical period... handled too casually.*

MANAGING TECHNOLOGY
DEPENDENT OPERATIONS

2. <u>Do not jump at every potential project</u>. Some projects can never return enough value to ITW to warrant the cost. An exception could be a small project done to further the working knowledge and relationship with the business unit. You might think of these as "advertising." Determine the potential value of the new project and prioritize it against your current projects. On average, we capture about 20 times the cost of the project for the business units, so you should consider the present value and costs of the project to meet this target. Obviously, you should not spend more than the potential present value of the project. Furthermore, you must be open and business-like in discussions about costs and benefits with the business units so that resources can be managed properly. Business units make cost and benefit decisions like these daily and will understand a decision not to proceed if it is made logically with good data and assumptions.

3. <u>Wait for new acquisitions to be integrated into the new corporate culture</u>. Many companies new to ITW are eager to use the Tech Center as soon as they find out about it. However, if projects are started too soon, you risk wasting the effort. As the ITW culture and business practices permeate the new company, the business unit's products and processes are evaluated from a new, perhaps more demanding perspective. This evaluation means that the target product of your project could be discontinued, the processes modernized to eliminate the problem, your engineer contact/project champion may be reassigned to higher-priority programs, or the champion may be let go in cost reductions. If a new business unit calls for a project early in their tenure, try to keep projects small and fast (Tech Service) until the items above are resolved. This could take as much as 1-2 years.

> *...Your project needs someone who leads the project on behalf of the business unit...*

4. <u>Get a champion</u>. Your project needs someone who leads the project on behalf of the business unit (a champion) at each level. The project must be clearly supported by the business unit's management, manufacturing, and marketing. For example, the plant engineer may be very excited and helpful,

but if his or her boss or other management has other priorities, the chances of success are low. Find the key people and learn their opinion. The champion must be truly enthusiastic, not a reluctant leader assigned by the boss. The project must be part of his high priority work. If the champion is not well regarded by the organization, the results of his work will not be trusted by management and your project will be second-guessed. The champion and the project need to have the full support of management. This means that the project should be the number 1 or 2 priority for the business unit. Otherwise, even if it is successful, it will probably not be implemented because the business unit will have higher priorities with limited resources.

5. <u>The most important champion is the customer</u>. A customer champion is critical for new products or major product upgrades, and of varying importance for cost reduction or manufacturing process projects. The most powerful predictor of project success is having strong communications with the customer. Communications that go through a sales representative and/or marketing manager may garble facts and prevent the dialogue needed for clear goal setting and prioritization of product features.

> *...insist on a direct meeting with the "outside" customer...*

You must insist on a direct meeting with the "outside" customer to make sure all requirements are known and proper tests are agreed upon. If there is no customer champion, or you are not permitted to visit or talk with them, consider killing the project now. A new-product project is almost certainly doomed without active communications with a customer.

6. <u>Hold detailed project discussions at the business unit before starting the project.</u> The items needed are: a set of clear performance, timing, and cost targets for completing the project; for determining the value of the project. This makes sure that all roles and responsibilities are clear to everyone at all levels of the business unit.

7. <u>Make sure resources are available to complete the project efficiently and rapidly.</u> Don't start until the proper equipment and manpower are

MANAGING TECHNOLOGY DEPENDENT OPERATIONS

> *Talk openly about resources and prioritization.*

available at both the Tech Center and the business unit. The business unit will usually have only enough staff to do one major project. If you have 3 or 4 projects going simultaneously with one business, some will not get implemented. Talk openly about resources and prioritization. If the level or type of expertise is not available, develop plans to build a team, find consultants, or get training for team members.

8. <u>Speed cannot be overemphasized.</u> The timeline of a project can be affected by market changes, competitors introducing new products, business strategy/tactics changes, or even by the economy stalling out or expanding. Also, personnel changes at the business unit can remove a key champion or put in place a new boss with different strategies. You have a much better chance to complete a successful project if you can finish in 10 months versus 30 months. You must plan for rapid progress.

> *...heavy project load that provides "comfort" - ensures slow progress...*

9. <u>Focus for speed</u>. If you have too many projects, none will progress fast enough. Make sure your current projects will not be delayed by adding a new one. It can be awkward to say "no" or "later" to a request for a project, but it will be easier than explaining delays later on. It is comforting for an engineer to have many projects thinking that if one fails, another will go forward. Ironically, the heavy project load that provides this "comfort" also ensures slow progress and more failed projects. Consider delaying project initiation until enough time is really available (i.e. current projects reach a "hold" phase, or are completed or killed). If the project need is pressing, you can consider building a team where another engineer takes on initial tasks, or have another engineer take on the project.

TECHNICAL SUPPORT TO MULTIPLE OPERATIONS
By James P. Nelson

10. <u>Start with the end in mind.</u> This is part of the teachings in Covey's book, *The Seven Habits of Effective People*. Talk about what represents a successful outcome, and stay focused.

WHEN OPENING A PROJECT

Ask the right questions— obtain the right answers: Having gone through the suggested steps before starting the project, there are follow-up considerations when a commitment is "formalized."

11. <u>Emphasize the Tech Center's need for project values.</u> Usually the marketing manager or business manager will be the best person to provide realistic project values. If a value cannot initially be estimated, getting that information should be one of the first project team tasks. It might be difficult to find the value for some projects. But if everyone agrees the project is worth doing, there will be a reasonable way to estimate value closely enough to make a decision whether to proceed. If you cannot get the business unit to give a value, maybe the project has, or such low priority that it is not worth their time to develop an estimate. Either low value or low priority predicts failure.

> *If you cannot get the business unit to give a value, maybe it is low priority...*

12. <u>Determine the business unit's commitment to the project.</u> If the business unit is not willing to share major expenses, machine time, etc., the project may not be worth pursuing. If the business unit does not assign staff, perhaps they see the project as low priority. The project must be part of the business unit's high-priority projects. If the project is not worth their time and money, it is not worth the Tech Center's time and money. A corollary is that no project should be pursued just because "engineering is free" from the Tech Center. There just has to be more value than that.

MANAGING TECHNOLOGY DEPENDENT OPERATIONS

13. <u>Identify the (outside) customer's stake in a new product</u>. What value is really delivered to the customer? Lower price is almost never a good reason for a program, because the competition is likely to reduce pricing and take away the margin for your new product. Also, customers will rarely pay more for incrementally new features or performance. Your product must be better <u>and</u> cheaper. If you assume the customer will pay more for new features, you will almost always be disappointed - maybe after spending a long time and a lot of money. Reducing or eliminating labor for cost reasons will seldom return the cost of engineering, development, cost of capital, and the cost of implementing a new process technology. In contrast, automating a process to increase product quality and consistency, or perhaps safety and ergonomics, will almost always pay off.

> *Your product must be better <u>and</u> cheaper!*

> *Beware of "product creep"... just as development is within reach of the original goals, Marketing or Manufacturing are tempted to propose new, "wouldn't it be nice" goals... leading to higher costs, delay, and chance for failure.*

14. <u>Brainstorming early in the process will provide many options for solving problems, reducing costs, and getting innovative ideas.</u> The key is to gather innovative staff with a very broad, collective experience. Bring in key players: retirees, consultants, and people from other business units. Consider using Tech Center staff that has been trained to lead brainstorming sessions.

15. <u>Set performance needs and try to get them "frozen" as early in the project as possible.</u> Marketing and Sales have a key role here. Being unfamiliar with product development, they may be tempted to go for very high performance, thinking it will make the sales job easy. What they may not realize, or forget, is that the development

time and cost may be several times higher than if they had goals that are more modest. Discuss the trade-off between highest performance, time, and money. Beware of "product creep" — sometimes, just as the development team is within reach of the original goals, marketing or manufacturing are tempted to propose new, more difficult "wouldn't it be nice" goals. Product creep will invariably lead to higher costs, delayed entry to the market, and increased chances for failure. A better option is to complete the version almost done. You have then captured the cash flow of "Phase 1" quickly.

16. <u>Keep asking, "How can we get to market faster?"</u> As mentioned above, managing project timing is critical. You will find a Gantt chart valuable to set timetables and responsibilities. The business unit will appreciate a business-like approach. But do not get carried away with too much detail. A Gantt chart is primarily a planning and communication tool, not the end product. The process of making one is perhaps more important than the final plan, as it helps all team members understand the complexity and issues everyone is facing. Do not try to do everything yourself; learning new skills as you go will delay the project. Look for a team member or consultant who already has the skills, let them proceed fast, and learn on the job.

17. <u>Look at the big picture of the business.</u> Discuss the project with the business unit's non-technical management, such as the General Manger or Vice President, to make sure they still support the project and share priorities. They will often make good contributions by providing critical business perspectives you cannot get from others. You are also likely to learn their concerns, perhaps developed because of problems on earlier programs. You can then address those in the normal course of the program, instead of learning them at the capital approval stage and delaying matters in order to get the answer needed for decision.

18. <u>Now is a good time to analyze the competition.</u> Find their patents, technical papers, and press releases. Estimate their cost structure and margins, product features, and distribution channels. Start a Tech Center support team for advice and help on technical issues outside your expertise.

MANAGING TECHNOLOGY
DEPENDENT OPERATIONS

19. <u>A new product project must have direction and leadership by marketing.</u> The entire team must have a clear view of the customer wants and needs. Direct and honest customer feedback is critical. How does the customer value and rank various product attributes (including price)? The project team must consider the time and cost to put certain features into the product and compare that time and cost with the value delivered to the customer. Here a technique called "Value Analysis" can guide decisions. Marketing has to estimate product volume and pricing, consider distribution channels, and convey that to the development team.

> *… if business balks at paying a share of the costs, you have a "red flag"…*

20. <u>Manage project costs.</u> Tech Center costs should be considered and scaled to the benefits to be reaped by the business unit. The same effort should not be made on a $100,000 project as a $10,000,000 project. Project development costs can easily exceed the budget of the Tech Center. The engineer/project manager must alert the business unit at the start that large expenses will probably require business unit support in part or in total. If the business balks at paying a share of the costs, you have a "red flag" that they really do not value this project highly.

21. <u>High capital costs can stall or kill a project.</u> If the project implementation requires capital expenditures large enough to warrant approval by higher management, then the team must learn the approval process and the criteria for approval. These factors will change as the general business outlook changes. In an environment with lower sales and earnings, getting approval for capital will be much harder, or impossible. The project leader must lead business managers to direct project work to meet corporate capital allocation goals. Start with the end in mind.

TECHNICAL SUPPORT TO MULTIPLE OPERATIONS
By James P. Nelson

MANAGING ONGOING PROJECTS

Ways to Keep Things Moving: Even great projects can get bogged down and slow to a crawl. Fight this malady with the techniques listed here.

22. <u>Work on some tasks simultaneously</u>. Working sequentially will usually slow progress. There is often a time when you will go faster with more people. But watch for communication issues as teams grow and members come and go.

23. <u>Construct an aggressive but realistic timeline/Gantt chart.</u> Remember to add a week every 4-6 weeks to have a buffer for unexpected delays. Illness, vacations, shipping delays, computer viruses, etc., will eventually happen in every project. This is especially of concern for those projects that last more than 4-6 months. Consider to outsource services for faster, cheaper results.

24. <u>Don't work alone, even if it seems possible.</u> A single engineer is unlikely to achieve the speed and quality that can be gained by a team in today's complex projects. Use the collective knowledge of the Tech Center, other divisions, or consultants. A teammate can give feedback on ideas (That is a lousy idea! How about this...?). The ITW Technology Resource Website provides a forum where any ITW employee can post questions on virtually any topic to over 2,000 experts.

> *Having your outside customer do all the field tests will soon wear out your welcome...*

25. <u>Decide whether to go in process order or to do the hardest step first.</u> If the most difficult technical problem is at the end of the process, work on that one first. If the hardest step can't be done, then you have not wasted time on the easy steps. There are examples of projects as old as 2 years that finally tackled the last, hardest step, and then failed. At that point so

MANAGING TECHNOLOGY
DEPENDENT OPERATIONS

much resource has been expended that it is harder to abandon the program. So the "Cost to Kill" tends to gain momentum as the project gets older.

26. <u>Perform failure testing; be brutal in abusing your product</u>. Customers are ingenious in abusing products. Make sure your tests reflect the reality of use. You will get what you test. Lab tests are good screening devices but cannot give a complete idea of performance in the field. Your customer must provide all relevant required tests. First, make arrangements for your own field tests. Having your outside customer do all the field tests will soon wear out your welcome, especially if there are many failures. Keep your failures private and approach your customer for validation only after you have every expectation that your prototype will meet and exceed expectations.

27. <u>Don't forget the Beta-test.</u> It is often impossible to test every aspect of a new product in the variety of ways your customers will use and abuse it. Work with the business unit to find 2-5 trusted customers who are willing to try out the new product. Resist the temptation to just go ahead with production in order to get revenues flowing fast. Recalls on a new product introduced prematurely will usually doom it in the marketplace.

28. <u>Make plans to transfer technology to the plant.</u> This will be less problematic if plant personnel have been team members from the start. If that has not been the case, meet soon with plant personnel from the plant manager to the shop foreman. They may discover problems you did not envision. Also, they will often solve problems easily.

29. <u>Keep in touch with the business unit manager or general manager</u>. Are they satisfied with the progress? Has the business environment changed? Is the team functioning well? This review might be at specific intervals of time, project cost (e.g. each $25M), or at the next gate for those using Stage-Gate project management tools, i.e. formal process of project input, review, revision and approval.

30. Understand the business, process, internal politics, etc. Visit the business unit often, sometimes just to "hang out," unless travel is prohibitive. Growing a personal relationship will help get the project through the occasional rough spot and make the job fun.

31. Change the Lineup. If the project seems to be "stuck", not progressing, the team may have the wrong chemistry. Some teams don't "click" even when individuals are talented. Consider handing the project off to another Tech Center engineer, or perhaps explore changing participants at the business unit.

RED FLAGS FROM PROBLEM PROJECTS

To fix or to kill: As the project develops, keep alert for problems. These can kill your project, or delay it so long that no one cares anymore. Recognize these killers quickly and take action. Often you will need help from Tech Center or business unit managers. Good relationships nurtured from the beginning will help the team get through tough periods.

32. Team communications are poor. Symptoms include no response to phone calls, faxes, emails, or reports; and no face-to-face meetings for an extended time; or key staff skipping the meetings. These are often silent clues that priorities have changed at the business unit.

> *To fix or to kill: Recognize the killers quickly and take action...*

33. Sometimes the business unit will lose interest in the project. They may be reluctant to say this outright because they know how much effort you have put forth on their behalf. They might hope that somehow the project gets killed anyway, without the painful discussion. Some signs that priorities have changed at the business unit that you should watch for are: samples sent to the business unit are not evaluated or are lost, no value is provided in spite of repeated reminders, the project champion is reassigned or has

higher priorities, you find yourself doing most of the work, the business unit cannot find the money for their share of the costs, or it seems you are more concerned about the project than the business unit. If they do not care about the program, why should you?

> *...the well-known "NIH" attitude.... is hard to fix, and "killing" the project could be the best action.*

34. <u>Watch for lack of cooperation at the business unit.</u> The business unit may not be willing, or simply may forget, to share key information (often market information). Perhaps they do not have full trust in the engineer or the Tech Center. Occasionally, a business unit engineer may see the Tech Center as a competitor, someone to take all the credit; or they may take the well-known "NIH" (<u>N</u>ot <u>I</u>nvented <u>H</u>ere) attitude. These can happen when senior management asks the Tech Center to "go over there and fix the problem." If this is perceived at the plant, the chances for cooperation will often decrease. This issue is hard to fix, and "killing" the project could be the best action.

35. <u>If there is no champion who is working closely with you at the business unit and pushing for progress, the chance for success is low</u>. This can be another symptom of lack of commitment by the business unit. Are they putting key staff and priorities elsewhere? If so, it is time to go. This issue should be discussed with business management to see if they can suggest alternate people for the champion role, but a new champion may not be available because of the limited personnel at the plants.

35. <u>Lack of a customer champion is the worst problem</u>. A customer champion should be brought in and managed by a good sales and marketing staff. Is there a customer demanding the new product? Does the product really bring high value to the customer? Or, has a competitor or alternate technology beaten you to market? Stay close to sales so that you communicate often with the customer champion.

TECHNICAL SUPPORT TO MULTIPLE OPERATIONS
By James P. Nelson

36. <u>Some problems are closer to home. One that never seems far away is the engineer who is over-committed to too many projects.</u> Symptoms of this malady are slow progress on several projects, rushing to get experiments done, or rushing to get something done before the monthly report is due to the business unit. It is hard to say "no" or "later" when you know you can solve the problem. If you find yourself in this predicament, review your project portfolio and decide whether to transfer a project to another engineer who has fewer projects, or see if you have another project that could be put on hold.

37. <u>Poor project parameters can make final implementation difficult. A key problem is high capital investment.</u> High capital raises the bar for the project team. The savings or new sales should pay out the capital in ½ to 2 years. If capital is high, it is critical to know the criteria for approval by upper management. Sometimes this is not known at the business unit, especially if it is new to the company or if the approval process has not been used recently. Payout estimates for new products are difficult, and uncertainty makes a decision to spend large capital difficult. In this case, look for ways to reduce capital risk such as used or rented equipment or contract manufacturing. A good Beta-test can solidify sales forecasts and build demand, making the decision to commit capital easier. If the capital needs are high, a quick evaluation could lead to killing the project before resources are spent.

> *... look for ways to reduce capital risk.*

38. <u>If costs and margin requirements lead to pricing which is equal or higher than competitive offerings, trouble lies ahead.</u> Customers will almost never pay more, even for a superior product. Claims to the contrary will usually be self-delusional to avoid the hard decision to "kill". Remember, many competitors do not have the same high margin requirements we do, or may have depreciated hardware that permits lower pricing, at least long enough to kill your new offering in the marketplace. It is a hard world out there.

MANAGING TECHNOLOGY
DEPENDENT OPERATIONS

ENDING PROJECTS

Completing, Killing, and Shelving Projects: As you have successfully negotiated the problems above, the end is in sight. So now you can just go ahead to the next project, right? Wrong! Let's finish properly and then go work with the next client.

> *Let's finish properly...*

39. <u>First, confirm that the business unit also considers the work done</u>. Also, reconfirm that the value of the project is unchanged. Get a copy of the first purchase order to document the sale of new products. As appropriate, write a short letter to the business unit confirming the project is closed, confirming the values and other key facts, such as formulas, procedures, recommendations, etc.

40. <u>Some projects are not successful, and they will be killed.</u> The project team will normally make this decision, but occasionally the business unit or Tech Center management will reach that conclusion. Confirm with the business unit that they concur. Then issue a brief report to make sure everyone knows why the work is stopping. Don't let dead projects live on hoping for some miracle to occur. Kill it now; it can be reopened later when circumstances are more favorable.

> *Don't let dead projects live on...*

41. <u>Clean up the files</u>. Compile all important documents, discard other materials, and make sure both personal and department files are complete. Document all formulas, designs, and recommendations; note disclosures and patent applications. Complete project management database forms if you use them.

IMPROVING RELATIONS WITH BUSINESS UNITS

A Win-Win: It is good to look at the work done and to be done from both sides. Consider the following issues:

TECHNICAL SUPPORT TO MULTIPLE OPERATIONS
By James P. Nelson

42. <u>Good relationships with the business unit speeds-up the project.</u> They also make it fun. This relationship is more complex than one might first think. The engineer should connect at several levels: champion/engineer, business-management, as well as on the plant floor. Most of us find the first and last natural and okay. We have to work harder to make and keep the business management connection. One technique is to make a 5-minute "report" to the business manager whenever you work at the plant. Include the plant engineers and give them as much credit as possible.

43. <u>Connect with the champion/ engineer.</u> Take them to lunch or dinner once in awhile, bring a Tech Center gift, or bring donuts to the plant or to meetings. Get to know this champion on a personal level. De-emphasize your degree, as it can make some people at the plant nervous. Emphasize joint planning. If a patent disclosure is submitted, make sure the business unit personnel are included and put their names first. You have many chances for a patent; they have few. Dress to make the client comfortable. Shed the tie and get dirty with everyone else. Do some tech service work to build the relationship. Just be there. Make sure project reports do not embarrass the business unit. Do not embellish; be accurate.

> *Emphasize joint planning... shed the tie and get dirty with everyone else.*

44. <u>Be open about Tech Center and project costs.</u> You and the business unit are responsible to manage ITW resources. Share your expectations and ask about theirs.

45. <u>Extra effort is needed to work efficiently with business units outside the U.S.</u> Differences in language, cultures, markets, suppliers, and pricing all lead to misunderstandings and delays. Know at least the basics of the local languages and culture. Some homework is required to avoid the "ugly American" impression. Find a project champion who is comfortable with the Tech Center, who can communicate easily, and who has time for the project.

MANAGING TECHNOLOGY
DEPENDENT OPERATIONS

45. It is important to start the overseas project with unhurried face-to-face meetings. Often it will take 3 or 4 of these to build a sound relationship. Following these with weekly teleconference calls is recommended. Video conferencing is becoming less costly and has high value, even for the difficult brainstorming stage. Web conferences can also help communications at low cost. However, the telephone is preferable, at least at the start.

> *...start the overseas project with unhurried face-to-face meetings...*

47. Identify the growth potential of the business where business units are small. Some are growing fast and will be a growing part of the Tech Center programs. Build a relationship early with projects that are fast and easy to complete. As the business grows, the projects will get bigger and have a higher value. Understand the product's pathway to the end user (1st tier? 2nd tier?). Cultivate internal and external links to make the project succeed. Visits to the customer/end user are very powerful in Asia because the business units are typically short of technical help and you can help in building a good image of the business unit, and of ITW. The connections and relationships will build slowly, but are very valuable. Support product demonstration to the business unit and its customer.

48. Explain how the Tech Center works and gives credit. Emphasize that the Tech Center does not duplicate or compete with the business unit. Make sure the business unit understands that we do almost no basic research. The Tech Center does not do turnkey work; the business unit's staff are the experts in the materials and processes. Successful projects require manpower and resources from the business unit.

49. Be aware of history between the business unit and the Tech Center. If there are difficult personalities, or a history of friction, get some advice from others who may have had success there. If some people feel offended, find out what caused the problem and look for ways to smooth things over. Do not inadvertently talk about old problems. If there are superegos, they

will need extra attention. If there is a good history of completed projects, try to connect with the plant staff and find out how they would like to work with the Tech Center.

50. <u>Sell the Tech Center.</u> Use the Tech Center brochures to explain our capabilities and resources, but be ready to deliver when the business unit comes to you for help.

For a suggested check-list of questions to consider when serving a business unit, customers from a centralized technical resource is given in the Appendix.

Suggested Additional Reading

1. *Product Development*, by Michael McGrath, Michael Anthony and Amram Shapiro

2. *The Gifted Boss*, by Dale Daughten

APPENDIX

Table A-1: Before Starting Projects:
Do not jump at every potential project- carefully choose and organize

- Have you visited the business (customer) often?
- Is the business a new acquisition (new customer)?
- The most important champion is the customer. Do you have a champion at the business unit?
- Did you have detailed project discussions before starting the project?
- Are all resources needed available?
- Speed cannot be overemphasized. Are you focused for speed?
- Do you have a clear vision of the end result?

Table A-2: When You Open a Project:
Ask the right questions— obtain the right answers

- Do you have a project value? What value is really delivered to the customer?
- Is the business unit really committed to the project? Does the customer have a strong stake in a new product?
- Did you set performance needs and get them "frozen"? Did you Brainstorm?
- Do you have a vision of the big picture of the business? Can you get to market faster?
- Did you analyze the competition?
- Is marketing providing direction and leadership?
- Are you managing project costs?
- Does the project need high capital?

TECHNICAL SUPPORT TO MULTIPLE OPERATIONS
By James P. Nelson

Table A-3: Managing Ongoing Projects: Ways to Keep Things Moving

- Can you work on some tasks simultaneously?
- Did you construct an aggressive but realistic timeline/Gantt chart?
- Is it a one person effort – are you working alone? How about changing the lineup?
- Should you do the hardest step first?
- Have you been brutal in abusing your product?
- Have you forgotten the Beta-test?
- Is the production plant (customer) up to speed?
- When is the last time you talked with the business unit manager or general manager?
- Are you clued into the business, process, internal politics, etc?

Table A-4: Red Flags for Problem Projects: To Fix or To Kill

- Are team communications poor? Is there good cooperation at the business unit?
- Has the business unit lost interest in the project? Is the champion still with you?
- Is the customer a champion?
- Do you have too many projects?
- Are there poor project parameters or high capital investment?
- Have you looked lately at costs, margins and pricing?

Table A-5: Ending Projects:
Completing, Killing, and Shelving Projects

- Does the business unit also considers the work done?
- Who will care if this project is killed?
- Two years from now, if you are gone, could the project be continued?
- Clean up the files. Are project closure forms done?

Table A-6: Improving Relations With Business Units / Customers:
A Win-Win

- Is there a good relationship with the business unit / customer?
- Are you well-connected with their champion/engineer?
- Are you open about Tech Center and project cost?
- Do you use special care with business units outside the US?
- Did you have unhurried face-to-face meetings?
- Did you explain how the Tech Center works and gets measured?
- Are you clued in to the history between the business unit and the Tech Center?

Chapter 11, Lead-in:

PROJECT CONTROL

One of the most difficult tasks in operations management is to introduce change. After an operation is made to run smoothly- establishes standards and standard practices, and becomes routine - to introduce a change takes considerable effort. It is easier when the organization itself is flexible and project-oriented. Some of the more strategic elements of project prioritization and project portfolio management have been discussed in Chapter 9. Chapter 10 provided some tips to project management as a part of a multiple operations support role.

Ultimately attention has to be given to the mechanics of control of each project by itself. This is the subject of this chapter. In Chapter 3, the reader was given an abstract overview of the basics of an Operation Management course in a possible MBA course. In this chapter the reader is provided with the basic tools of project management as could be expected from a Project Management Seminar. It should be useful to the non-technical, general manager to grasp what are the "standard" tools of project management. Given appropriate circumstances the manager may wish to recommend the use of any of referenced software as the situation demands.

CHAPTER 11

PROJECT CONTROL

Eric A. Spanitz

The secret to project success is having a written goal and scope. In fact, by writing down the goal of the project, you have a 100% better chance of success.

In the first two sentences of this chapter, you just learned the secret to successful projects! The intent of this chapter is to guide the reader through the knowledge necessary to effectively define, plan, and manage projects – all as an executive level overview. Please refer to the appendix for additional resources to further your exploration of project management best practices.

> *...managed control of activities...*

Project management is the planning and managed control of activities to accomplish some defined end-result, within the defined constraints. Therefore, project management, in some form or another, has been around for thousands of years. What is generally considered "formal project management" started with the U.S. war efforts during World War II. Project management as a modern profession is said to have been started with the founding of the Project Management Institute.

The **Project Management Institute** is an international organization that has been around since 1969. With almost 90,000 members worldwide,

PROJECT CONTROL
By Eric A. Spanitz

PMI® is the leading nonprofit professional association in the area of Project Management. PMI has collected and recorded best practices, conducted research, and has furthered the standardization of project management practices around the world. The PMI publishes a booklet that serves as a collection of standard project management concepts, at a very cursory level. This document is known as "A Guide to the Project Management Body of Knowledge" or PMBOKGuide®.

> *The **PMBOK**® is a ... "concept dictionary" for project management...*

The **PMBOK GUIDE®** is a summary document that describes the "sum of knowledge within the profession of project management" (according to the PMI). This one booklet provides a brief and basic summary of the many papers, articles, and books published by the PMI. Some in the project management profession refer to this as a "concept dictionary" for project management. The PMBOKGuide® was first published in 1987, then updated in 1996, and more recently in 2000. This document is updated by a group of committees, and must go through an international ratification by the members of the PMI. The PMBOKGuide® is considered the formal and final say in project management definitions and concepts. Some new to project management think that they can read the PMBOKGuide® to learn how to practice project management, however that is like saying a quick read of a dictionary will teach you how to speak a language.

According to the PMBOKGuide®, a **project** is "a temporary endeavor undertaken to create a unique product or service." Breaking this definition down to the essential core components, a project has the following elements:
- Beginning – a defined and known starting point
- End – a clearly defined ending point
- Defined Goal – a measurable, quantifiable end-product
- Tasks – activities that are steps towards completion
- Resources – anything that helps complete a task

MANAGING TECHNOLOGY DEPENDENT OPERATIONS

> *...a project that is in trouble is one where those involved cannot articulate*

Note that a key element of a project is its defined starting point and defined ending point. One sign of a project that is in trouble is one where those involved cannot articulate when it (supposedly) started, or when it is supposed to finish. The PMI also advocates that any project's overall duration should be less-than 1.5 years. Any project longer than that should be broken up into smaller pieces, simplifying the planning and managing.

A program is a collection of related projects that accomplish an overall goal. So an endeavor lasting over 1.5 years should be known as a *program*, made up of shorter projects. While this might seem like an exercise in semantics, the author wants to ensure the reader uses these formal project management terms properly. This is especially true considering the existence of the widely referenced PMBOKGuide®.

So what about activities that are ongoing, operational in nature, and without a defined end? Those activities, such as payroll, are a **process**. While a project attempts to satisfy a defined and unique need, a process is more closely associated with "business as usual," or day-to-day work.

As was already mentioned earlier, a project is made up of five basic elements: a beginning, an end, a defined goal, tasks, and resources. There are also four recognized limits, or boundaries, within which any project must operate. The four main constraints of any project are: Time, Resources, Goal-Scope-Quality, and Morale. These four main project constraints are usually represented as a triangle or known as the "Triple Constraints." Yes, there are four of these "triples" so please read on to find out how that happened…

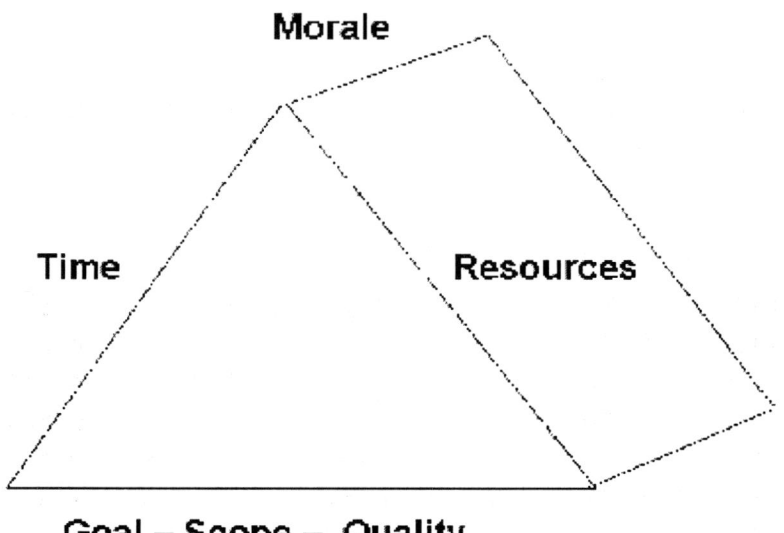

PROJECT CONTROL
By Eric A. Spanitz

```
            Morale
           /\
          /  \
         /    \
    Time/      \Resources
       /        \
      /          \
     /_____\
   Goal – Scope – Quality
```

Time refers to the duration available to complete the overall project. Most inexperienced project managers think that this constraint is the most restricting, however, as will be discussed later, this is actually the constraint that tends to handle itself. Regardless of efforts (or lack thereof) time will, indeed, pass...

> *... the constraint that tends to handle itself.*

Resources are anything that helps work on activities in the project. Resources can be people, machines, licenses, or even piles of money...anything that helps complete an activity.

The **Goal** refers to the measurable and quantifiable end-result of the overall project. This is ultimately what the project is trying to accomplish. Directly related to this is the **Scope**, which refers to the boundaries or limits put on the goal. Considering **Quality** as part of the project goal emphasizes how project stakeholder expectations must be defined as part of the goal to ensure both project success and customer satisfaction. In other words, quality is an

> *...deliver quality ...and manage customer expectations.*

inherent aspect of the goal and must be defined as meticulously as the product the project is to produce. One generally accepted definition of quality is "meeting or exceeding customer expectations." The only way a project can meet or exceed customer expectations, and therefore deliver quality, is to discover, record, and manage customer expectations.

Morale is the fourth of the Triple Constraints and is included as a fourth dimension in order to emphasize the important influence of morale on project success. A project is more than just a list of activities and deadlines. It is the people working on the project that will make – or break – the project.

Why the triangle? The triangle represents the automatic trade-offs that happen if one of the constraints shift. For example, if the project must suddenly be finished quicker (the time side shortened), either more resources will have to be added (the resource side lengthened) and/or the Goal of the project will have to be pared down (the goal side shortened). Regardless of what happens to the three other sides, the Morale side will always be affected. This four-sided triangle is a very useful graphic to use when discussing project constraints.

We can now compare our earlier definition of project management with the official definition from the PMBOKGuide®: "The application of knowledge, skills, tools, and techniques to project activities in order to meet or exceed stakeholder needs and expectations from a project." Simply put, project management is the balancing of the project constraints while accomplishing the project's goal.

In the beginning…

When first starting a project, the overall project goal must be defined. The document used to define the project is sometimes known as a **Project Charter**. The project charter is also known as the "memo of understanding" or "statement of work." This project charter lists the mandatory and discretionary components of the project, sometimes known as the "must haves" and the "nice to haves." It is crucial for proper management of the project, that these two categories of deliverables are clearly defined; otherwise it is not possible to appropriately adjust the project constraints.

When defining the goal of the project, the goal must be **measurable** and **quantifiable**. This is key to the success of any project endeavor; otherwise the completion of the project can end up being argued between the opinions of two or more people. By avoiding words ending in "ness" (e.g. happiness) and "tion" (e.g. satisfaction), a project manager can help guide the definition of the project goal *towards* objective measures and *away* from subjective measures.

> *...the goal must be **measurable** and **quantifiable***

At the start of this chapter, you learned the secret to project success: a written goal. A written goal forces the project stakeholders to focus their efforts, as well as clarifies what will and will not be done. In fact, a useful document to create as part of the project definition is the **Not List**. The Not List is (literally) a list of things that will not be part of this project. This helps to define the scope, or the limits on the project. Experienced project managers put this document in front of all other project definition documents to make sure stakeholders and team members all see what is not to be assumed to be part of the project.

Another very useful scope-management document is the **Assumptions List**. This document is not just a collection of assumptions (e.g. "The project team will have timely access to the project sponsor"); it is also used to help document and clarify the reasoning behind the decisions made about the project plan (e.g. "The project team will use Outlook as a scheduling program because of its existing prevalence in the organization.").

Some organizations emphasize the need for an **Executive Summary**, which explains how the project fits in with the mission, objectives, and needs of the organization. This document can be important in clarifying the executive upper-level support and financial backing for the project.

MANAGING TECHNOLOGY
DEPENDENT OPERATIONS

> *Executive Summary... a wonderful internal communication document!*

The Executive Summary, combined with the Project Charter, and then condensed down to a brief summary no more than one page in length is known as the **One Page Press Release**. This "press release" is not literally a press release; it is a summary document that lists the most important elements, resources, and constraints of the project. This document serves many purposes: it helps the project manager focus his or her understanding of what the project is ultimately about, it is a wonderful internal communication document, and when distributed to contractors and vendors, it helps them understand what the project is about, without giving too many details (or confidential information) about the project.

The following table shows the commonly used project definition documents and their customary order of presentation:

One Page Press Release
Not List
Assumption List
Executive Summary
Project Charter – Statement of Work
...the rest of the project documentation...

Fail to plan? Then plan to fail...

Once the project goal has been defined, work can begin on planning the project. Keep in mind that project planning is not a one-way step-by-step process. Projects are planned and the project plans are adjusted the same way good business proposal should be created: through an iterative and increasingly refined combing back through existing documentation. What this means is, as work begins on the project plan, the project goal should be reviewed and fine-tuned as necessary. As additional work is put into the

PROJECT CONTROL
By Eric A. Spanitz

project documentation, the Not List and Assumption List should be updated. This review should be done to all project documents, throughout the life of the project.

Keeping in mind the document review process, one of the first steps in *planning* a project (as compared to *defining* a project as was just discussed) is to list the major deliverables for the project. Common practice is to use sticky-notes to list one deliverable per note, and stick the notes up on the wall. The deliverables can then be grouped by common elements, resources needed to complete them, or some other logical method. With the notes lined up on the wall, more detailed activities can be listed on additional sticky-notes and stuck beneath the major deliverables that they comprise. What is being described here is a **Work Breakdown Structure** (WBS). A sample WBS is shown below.

Figure 1: Work Breakdown Structure

```
                    Spanitz Haus
         ┌──────────────┼──────────────┐
     Foundation        Walls           Roof
      ─Clear site      ─Purchase wood  ─Lay tar paper
      ─Dig hole        ─Purchase nails ─Afix shingles
      ─Brace hole      ─Build frame    ─Attach gutters
      ─Pour cement
```

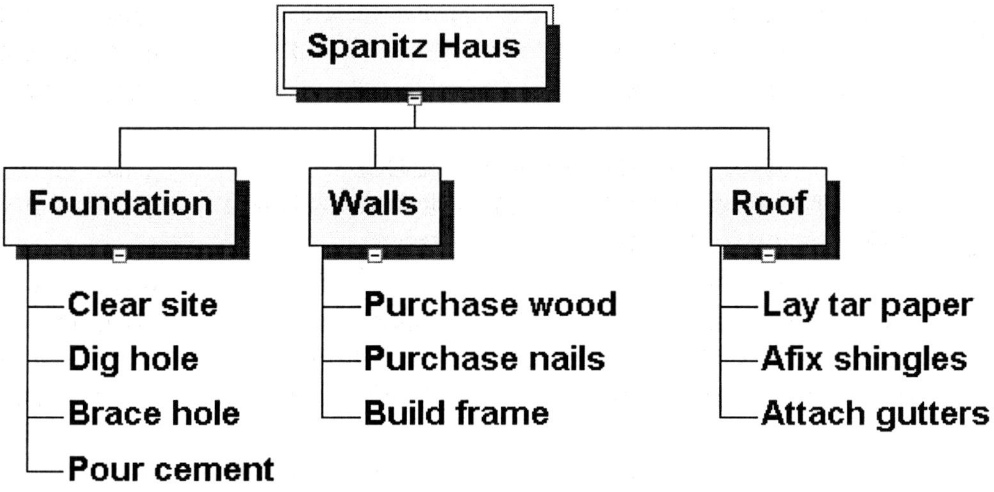

A WBS is used to give structure to the brainstorming process. It helps encourage the completeness of planning, rather than a sequential planning process. The PMI actually encourages this type of planning over what is known as the "3D" approach, which stands for "details, durations, and dependencies." The PMI recognizes that the 3D approach encourages a lock-step thinking process and leads to over-planning the beginning of the

MANAGING TECHNOLOGY
DEPENDENT OPERATIONS

> *...a lock-step thinking process ... leads to over-planning.*

project and under-planning the middle and end of the project. The WBS process, however, encourages a "fill-in" approach, which leads to a more balanced and complete first attempt at brainstorming project activities. As a bonus, the WBS can be created by several people at the same time, all posting notes onto the wall in the same room. This group-brainstorming process not only encourages the realization of activity interdependencies, but also can be a subtle team-building activity for stakeholders and team members.

The project WBS is usually then turned into an outline format by entering the tasks into a project scheduling package, such as Microsoft Project. This default outline format shows task names on the left and graphical bars on the right. The bars indicate each task's start and finish dates, as well as task duration. This type of a chart is known as a **Gantt chart**. A sample Gantt chart is shown below.

Figure 2: Gantt Chart

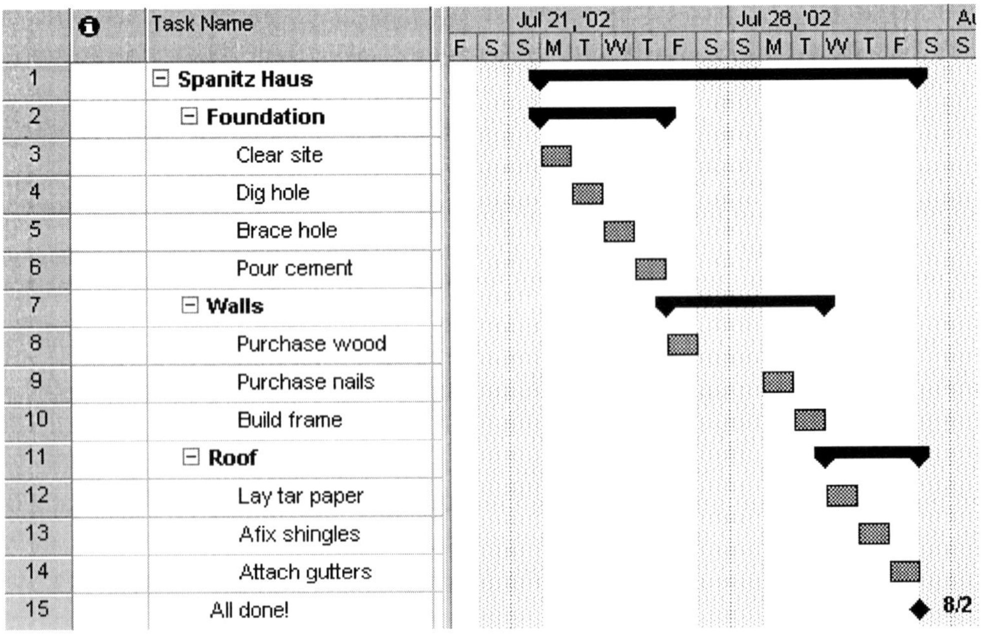

#	Task Name
1	⊟ Spanitz Haus
2	⊟ Foundation
3	Clear site
4	Dig hole
5	Brace hole
6	Pour cement
7	⊟ Walls
8	Purchase wood
9	Purchase nails
10	Build frame
11	⊟ Roof
12	Lay tar paper
13	Afix shingles
14	Attach gutters
15	All done!

Note that this type of organization of project data shows the activities on a calendar of sorts (note the dates on the right side of the chart, above the bars). Traditionally, the tasks listed towards the top of the Gantt chart happen towards the beginning of the project, and those towards the bottom of the Gantt chart happen towards the end of the project. This is not a set rule by any means, but is encouraged to make the Gantt chart easier to read and navigate.

Ideally, the Gantt chart should include tasks with zero duration, which serve as checkpoints in the project plan. These checkpoints are known as **Milestones**, and are traditionally represented by a diamond on the right (graphical) side of the Gantt chart (note the last line of the above Gantt chart labeled "All done!"). These milestones assist the project manager in tracking of progress throughout the project by giving a clear visual indicator of the completion of major project elements.

Working within the Gantt chart, task estimates should be entered. The estimates should be initially based on duration, as if this activity can be worked on without interruption. Later in the planning process it will be factored in that rarely do activities happen uninterrupted.

> ...*rarely do activities happen uninterrupted.*

Estimates vs. Guestimates

There are four sources of an estimate: (1) the historical experiences of the person doing the estimating, (2) the experiences of someone else, (3) the historical *similar* experiences of the person doing the estimating, or (4) the *similar* experiences of someone else. A reasonable estimate starts with a comparison to what has happened before. If there is truly no comparison, then the activity must be broken up into finer detail, to the point where a comparison can be made. This breaking up of the task into finer detail is the first of six tips for better estimating:

MANAGING TECHNOLOGY
DEPENDENT OPERATIONS

1. Break the task into finer detail – this encourages a more exact comparison to previous work activities and makes the item being estimated more manageable.

2. Add up the pieces – when breaking things into finer detail, the tendency is to slightly pad the item being estimated. There is also the concern of rounding-error. By adding up the pieces, the estimator can verify the feasibility of the estimate.

> *A reasonable estimate starts with a comparison to what has happened before.*

3. Use a point of reference – this is another reminder to look for exact matches with previous experience as well as similar experiences. With physical estimates, sometimes this means a literal point of reference is used, such as a measuring device.

4. Use consistent units – while the project scheduling software can usually auto-translate different units, it is easier for the humans to stick with one consistent unit. This also makes comparisons across projects easier and more consistent.

5. Ensure everyone has a clear definition of what is being estimated – probably the most significant influence on estimates is the definition and understanding of what it is that is being estimated.

6. Be aware of personal bias – as was pointed out earlier, the source of any estimate is someone's previous experience. This must be tempered with the existing situation and constraints, so that historical values do not automatically become "the estimate" for the current situation.

One of the most commonly used estimating techniques is called **Forecasting**. Forecasting is when the just-mentioned six tips for better estimates are considered, as well as a formal listing of what variables could affect the situation this time around.

An example of forecasting: the last time Pat wrote a report, it took two days. This time around it is close to the holidays, Pat is mentoring a new employee, and the task itself involves a new process the organization just

PROJECT CONTROL
By Eric A. Spanitz

adopted. Factoring in these three situation-specific variables, an estimate of four days will be recorded instead of the two days it took last time. There is no special scientific formula used here; this adjustment is dependent on the judgment of the estimator.

"Organizing is what you do before you do something, so that when you do it, it's not all mixed up." – A.A. Milne

After determining the durations of the activities, the planner must determine the sequence of events. This is usually the second or third pass at this particular activity, because usually when the WBS is transferred into the Gantt chart, the activities are put into some sort of order. As people continue to give their input into the project plan, the task order is further refined. When using project-scheduling software, this element of project planning is called **Setting Dependencies**. Setting Dependencies not only involves the sequencing of events, but also involves the determining of the type of connection between events. Do the tasks both start at the same time? Should the tasks finish at the same time? Will the delay of one task affect the timing of another task? These questions must be addressed during this part of the project planning. The knowledge required to make such judgment calls relies on the experience of the person setting the dependencies.

> *Should the tasks finish at the same time?*

If the planner does not have the necessary experience or knowledge, he or she must track down a **Subject Matter Expert** (SME) who does possess the correct experience and knowledge. A SME is someone who assists in the project planning, but does not actually get involved in the work. Think of an SME as a living reference book; someone whose experience can bolster the knowledge of those working on the project.

Once the sequence of tasks has been determined, the project manager should determine the **Critical Path** for the project. According to the PMBOKGuide®, the Critical Path is "the series of activities that determines

MANAGING TECHNOLOGY
DEPENDENT OPERATIONS

> *...Critical Path determines the duration of the project.*

the duration of the project – in a deterministic model, it is those tasks with float less-than or equal to a specified value, often zero." Tasks on the Critical Path are those activities that directly control how early a project can be finished. The reason it is important to determine the Critical Path is because these are the activities that should be reevaluated, adjusted, or compressed in order to shorten the overall project duration.

Here is a simple example of the Critical Path: if the project is to cook Thanksgiving dinner, including cooking a turkey, squash, and corn, which of the items determines when everyone can sit down for dinner? The answer is the cooking of the turkey because that activity takes the longest. Cooking the corn quicker will not speed up the overall project. Nor will cooking the squash any faster. Because the cooking of the turkey takes several hours longer than the other cooking activities, it directly controls the end of the project (when everyone can eat).

The next graphic is known as a **PERT Chart**, and is also used to show dependencies and relationships between project activities.

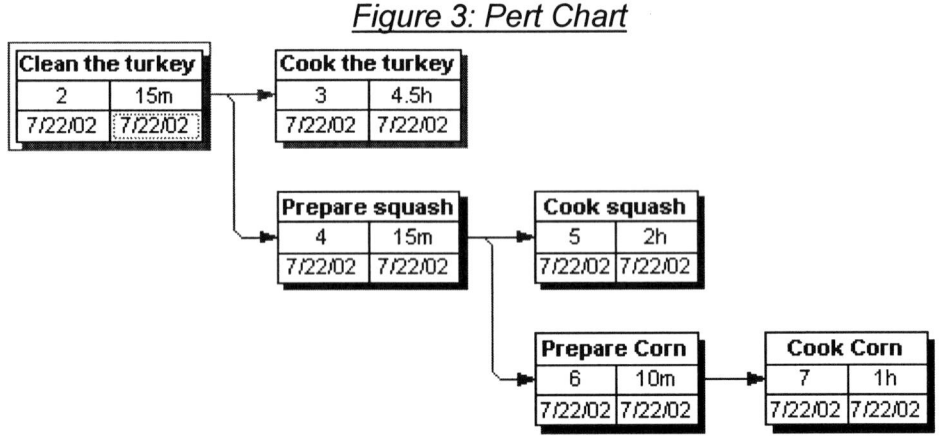

Figure 3: Pert Chart

Clean the turkey			Cook the turkey	
2	15m		3	4.5h
7/22/02	7/22/02		7/22/02	7/22/02

Prepare squash			Cook squash	
4	15m		5	2h
7/22/02	7/22/02		7/22/02	7/22/02

Prepare Corn			Cook Corn	
6	10m		7	1h
7/22/02	7/22/02		7/22/02	7/22/02

Note that it is not immediately obvious that the cook turkey activity is taking such a long time to complete, that it is controlling the end of the project. Next is a Gantt chart for this project, giving a reader a comparison between these two types of project charts.

PROJECT CONTROL
By Eric A. Spanitz

Task Name	Duration
⊟ **Thanksgiving Dinner**	**0.59 days**
Clean the turkey	15 mins
Cook the turkey	4.5 hrs
Prepare squash	15 mins
Cook squash	2 hrs
Prepare Corn	10 mins
Cook Corn	1 hr

Note how with a modern usage Gantt Chart (with dependency lines displayed) both the task durations and the dependencies can be depicted in a manner that makes Critical Path identification much easier. In fact, through the use of formatting and highlighting features, a user of project scheduling software can even have the software automatically colorize the Critical Path.

Availability is not a skill set.

As was mentioned earlier, resources are anything that helps you finish a task. Almost always, resources have limits. A project must take these limits into consideration; otherwise the project plan is only a fantasy. This brief section about resource availability focuses on the three main resource limits: maximum project availability, number of effective concurrent tasks, and overtime's diminishing returns.

A resource's **Maximum Project Availability** is the amount of time, usually expressed as a percentage, that a resource can work on project work, verses day-to-day duties. Research done by the PMI, universities, and several large corporations that did not believe the original findings all find the upper limit for maximum project availability to be 65%. In fact, more recently the University of Michigan repeated the study in May of 1999 to find that maximum project availability is now 60%. So what is happening with the other 40% of a resource's time? The key is remembering and factoring in those day-to-day activities that must take place but are not accounted for in the project plan as a line-item task. Consider as part of these day-to-day activities: coffee breaks, bathroom breaks, walking to the printer,

MANAGING TECHNOLOGY
DENDENT OPERATIONS

> *... remembering those day-to-day activities that must take place, but are not accounted for ...*

photocopying, meetings, phone calls, more meetings, e-mail, and even more meetings. In fact, the depressing part about this statistic is that the measured 60% maximum project availability took place with the subjects knowing their behavior was being monitored and recorded, and the results being provided to their management. One would expect that the subjects were hustling more than not during these studies – so what about when resources are not being directly monitored?

Remember that expecting a maximum project availability percentage higher than what is possible only ensures that all estimates about duration and work are overly optimistic. It is actually safer to assume lower maximum project availability, and finish tasks ahead of schedule.

Project scheduling software can help the user monitor a resource's maximum project availability, but another resource limit, the number of **Effective Concurrent Tasks** is more specific to the individual resource, and can be more difficult to determine. The Effective Concurrent Tasks is the number of activities a resource can effectively switch between (or work on "at the same time") before the work being done on these activities degrades. Research done at Indiana University has shown that the maximum number of Effective Concurrent Tasks is six, with most of the population having a maximum of four, and a large chunk of the population only able to work on one activity at a time. Remember that the word "effective" is important in this consideration. People can work jump between many activities, but at some point their efforts either degrade into poor performance on many (if not all) of the activities, or the individuals spend more time attempting to managing the juggling of activities than on the activity work itself. How does one determine this maximum? The simplest method is by asking the individuals themselves. Another (slightly meaner) approach is to continue assigning additional concurrent work until there is a noticeable decrease in the quality or performance.

PROJECT CONTROL
By Eric A. Spanitz

The third and final resource limit is **Overtime's Diminishing Returns**. A generally accepted project management rule of thumb is that for the first hour over and above the usual and standard work day results in approximately sixty minutes worth of work being done. The second hour results in approximately forty minutes worth of work being done. Please refer to the table below:

First hour	~ 60 minutes
Second hour	~ 40 minutes
Third hour	~ 17 minutes
Fourth hour	~ 7 minutes

Note that after the second hour, a lengthy break should be taken, or the person should just stop and go home! In what first appears to be a surprisingly compassionate statement, the PMI has said that it is "extremely bad technique" to plan overtime into the project schedule. This is not a pro-morale statement from the PMI; it is just a matter-of-fact acknowledgement that planning or expecting overtime takes the flexibility out of the project plan, which makes the plan less feasible, and also eliminates one of the more effective ways to handle the unexpected.

Waste happens.

A good plan is one that has not neglected to consider any crucial elements of the project. A better plan is one that has considerations built into it, to handle both the expected and the unexpected. A perfect plan does not exist.

According to the PMBOKGuide®, project risk is "an uncertain event or condition that, if it occurs, has a positive or negative effect on a project objective." There are two overall categories of project risk: known/unknowns (k/unk) and unknown/unknowns (unk/unk).

> *A perfect plan does not exist!!*

MANAGING TECHNOLOGY
DEPENDENT OPERATIONS

Known/unknowns are those risks that are acknowledged, but it is not known when or how bad they will be. An example of a known/unknown is one of the necessary resources calling in sick. This is a distinct possibility, however who and when and for how long, is not known.

Unknown/unknowns are those risks that the project personnel did not even know were a potential risk (or sometimes, did not even know were possible!). An example of this would be...well...not possible. As soon as a risk is identified, it is no longer in the category of unknown/unknown. You can know what you don't know, but it is also possible to not know what you don't know.

So how can someone handle risks that cannot even be identified by their very nature? Project flexibility is the answer. **Project flexibility** is having flexibility or negotiation room in one or more of the main project constraints (time, resources, goal-scope-quality, or morale). To plan a project so that there is no flexibility in the project constraints means the slightest delay, or the most minor unexpected event, will derail the project. This instability of the project plan, in most environments, is unacceptable.

It is the responsibility of the project manager and the organization's management team, to make sure that each project plan has several areas with project flexibility. Flexibility with the project budget is commonly known as a **Contingency Reserve**. A Contingency Reserve is a set-aside amount of money that is held back by the project manager in case (or when!) some of the monetary estimates are a little short. One of the most common ways this money is set aside is for the project manager to take the entire project budget, subtract five to ten percent, and only plan with the remaining amount of money.

A **Management Reserve** is a set-aside amount of money that is held back by the organization's management team for use *across* projects. The money is not under the control of the project manager or project team, and petitioning for its use usually involves a formal review of the project. Money

PROJECT CONTROL
By Eric A. Spanitz

used from this reserve takes money away from other organizational endeavors.

There are several ways "timing flexibility" can be added to a project. The project calendar can be made to be more conservative. This can take the form of a shorter "project day" (with fewer hours expected of work each day than people actually work) or additional half-day holidays,to take into account those days when people leave early en masse (the afternoon before an official holiday). Remember, this is not padding the schedule; it is making it more realistic!

Goal and scope flexibility is inherent in the way the goal and scope should be defined in every project. Remember that as part of the goal, project personnel must identify what is necessary and what is "nice to have." These mandatory and discretionary goals help add flexibility to the plan should the need arise. It is important to remind project stakeholders that a discretionary (or "nice to have") goal can be shifted to a follow-on project, completed after the main project is completed, or can be elevated to the level of a project in its own right.

> *...identify what is necessary and what is "nice to have."*

Morale can be the most difficult constraint to add flexibility to, however, it can also be the easiest. By proactively managing morale (e.g. treating people nicely before you need something from them, considering their personal work preferences, including them in the decision-making process), project personnel, contractors, and even project stakeholders can be encouraged to be more flexible with their work, demands, and attitude.

"The information you have is not the information you want.
The information you want is not the information you need.
The information you need is not the information you can obtain.
The information you can obtain costs more than you want to pay." –
unknown

MANAGING TECHNOLOGY DEPENDENT OPERATIONS

Once a project is properly planned the managing of the project can become almost automatic.

However, the only way a project can be managed and tracked is if it has been carefully planned. To paraphrase the Mad Hatter in Alice in Wonderland: *if you don't know where you are going, it doesn't matter much where you go.*

Managing a project involves four main elements: gathering information, analyzing the information, updating the project plan, and communicating about the project.

Gathering information can involve several methods: meetings, Management By Walking Around (MBWA), telephone calls, voice-mail, paper forms, faxes, e-mail, and software. Each of these methods has their own advantages and disadvantages, and all can be effectively used. There are two rules for gathering information:

1. *If you can't prove the use of the information, you are not allowed to ask for it.*
2. *Consider each person's "Preferred Method of Communication."*

If the project personnel cannot see the value or the use of the information collected, they will eventually stop supplying the information. Or even more damaging, they might start supplying false information. To avoid this, limit the information collected to that which is necessary to control the project. Do not collect information just because that particular piece of information is easy to collect! The project manager must also make conscious effort to clearly show (or "prove") that the information is constantly used to better control the project.

> *If the project personnel cannot see the use of the information collected, they will stop supplying the information.*

368

PROJECT CONTROL
By Eric A. Spanitz

One very effective method to prove the use of the information is to enter the data into the project scheduling software at the status meetings as people are reporting their progress. This effort must be managed properly; otherwise the meeting can degrade into several people watching the data entry person fumbling with the software. Appoint a scheduling software jockey to enter the data as the project manager runs the meeting. Towards the end of the meeting, show the project personnel the current Gantt chart, and point out the changes. Also mention where adjustments need to be made to keep things on track.

Another tip for collecting information is to use the 0-100 method. The **0-100** method uses a binary method of tracking task completion: either the task is finished (100% complete) or not finished (0% complete). This eliminates the ambiguity of having tasks hover around 80% complete for extended periods. The 0-100 method also forces the project planners to schedule short-duration tasks (usually one week or less) which in turn simplifies the defining of the task. Shorter duration tasks also simplify the resource assignments, as well as the tracking of project progress.

Each individual has a **Preferred Method of Communication**. This means they respond more positively and willingly to a particular method of communication. To facilitate the gathering of information, a project manager must balance the efficiency of standardized methods with the individual preferred methods of communication. This is, indeed, a delicate balancing act – one that cannot be eliminated by an attempt to force a communication method on unwilling project personnel.

> *...do not attempt to force a communication method on unwilling project personnel.*

When analyzing project information, do not lose focus on the overall big picture. Less experienced project managers tend to manage individual activities. This leads to micro-management. Good project managers "**manage to the milestone**." This managing to the milestone is a reminder

MANAGING TECHNOLOGY DEPENDENT OPERATIONS

that project personnel should always keep the overall goal in mind, and not obsess on minor schedule variances. This is another balancing act – one where the project manager must determine when a variance is an incident, and when a variance is a trend. While keeping an eye on the Critical Path can point the project manager in the right direction, distinguishing between an incident and a trend can only come from experience.

Project updates should happen frequently. As was mentioned earlier, project personnel are strongly encouraged to update the project schedule right as the information is collected. Some project teams even go so far as to have the project personnel update their own task data themselves. This, of course, introduces additional risks and coordination issues.

How often is frequently? Depending on the size of the project, the speed at which work is performed, and the number of activities, project updates can happen anywhere from hourly to daily, to even weekly or every other week. Remember that project documentation is living, and must be cared for as such. The authors know of one new project manager who reminded herself to update her project plan every time she watered the plant on her desk. What an interesting parallel to the living project documentation!

When reporting project information, there are two final reminders: be consistent and focus on how the project will be kept on track. Be consistent refers to the method, frequency, and form the reporting should take. Pick or design a report format, then do not continue to change it as the project progresses. Remember that any ideas for a better report should be recorded for the next project.

Focus on how the project will be kept on track means the project reporting should not just be a wall of words or a bucket of numbers. Project reporting should not just list all of the things that have gone wrong, nor should it just list the activities that have been finished. If corrections, adjustments, or changes are made, certainly they need to be mentioned. However the communication must remain focused on the project end result and constraints, and how these will be managed. Project reporting must communicate how a project is being guided towards successful completion.

"Efficiency is doing things right. Effectiveness doing the right things." – Zig Ziglar

Imagine that you have been selected to plan an Inuit wedding. You do your best and create a plan that seems (in your mind) to be pretty solid. You hand this project plan over to someone else to run, and then are instructed to plan another Inuit wedding. Do you see any potential problems with this situation?

Without watching and recording how a project plan actually evolves, the project participants run the risk of repeating their mistakes and not repeating techniques that seem to work. A review of a project, after its completion, helps project personnel and stakeholders identify what worked well, as well as ("perpetuates") what should never be repeated again ("prevents"). Commonly this formal project review is known as a **Post Project Review** or **Project Close-out**. These two terms are preferred over the out-of-fashion "post mortem" (which is Latin for "after death").

> *A Project Close-out should not be a blame session, nor should it be a product review...*

There are a few best practices for Project Close-outs. They should be held off-site, with food, last no more than an hour, and focus on what went well and what did not go so well. A Project Close-out should not be a blame session, nor should it be a product review (a product review should be held separately).By holding the Project Close-out off-site and with food, an environment is set that encourages honest discussion. Being off-site physically and psychologically distances the project personnel from their bosses and project stakeholders, which can foster a more open exploration of what not only went well, but also what might have been a mistake. The

MANAGING TECHNOLOGY DEPENDENT OPERATIONS

presence of food also shows appreciation by the organization for the project personnel's efforts. The one-hour limit helps keep the focus, and shows respect for the project personnel's time. And lastly, by preventing the meeting from degrading into a blame session, a project manager can cull better Lessons Learned both as perpetuates (should repeat) and prevents (must never repeat).

So by writing down a project's perpetuates and prevents, project personnel have a better chance of success in their upcoming project endeavors. This brings this chapter full circle: by writing down a completed project's good and bad practices, considering them, and writing down the goal of the new project, a project is almost guaranteed to be a success. Well, maybe luck still factors in…

Appendix 1
Project Management Software Applications

This list of project management software applications highlights the major players in the industry, as well as those applications that are must-haves. The reader is cautioned to invest in training on these software packages. "Winging it" with a project management package will only bring the user pain and sorrow.

Microsoft Project – by Microsoft Corporation
www.microsoft.com/office/project/

With over 98% of the market share, this program is the default standard in project management software. While some would argue that it is not the best software, it seems to allow users to get by with project schedules that are work "good enough." Unfortunately, this flexibility is also how most users misuse this program, then wonder why their schedules do not work as expected.

Primavera Project Planner (P3) – by Primavera Systems, Inc.
www.primavera.com/products/p3.html

This is the Cadillac of project management software...and costs almost as much. With the numerous ways this software integrates into an organization's operational software, this is the most powerful of scheduling packages. It is also the most expensive and by far the most complicated.

SureTrak Project Manager – by Primavera Systems, Inc.
www.primavera.com/products/sure.html

SureTrak Project Manager is the "baby brother" of Primavera Project Planner. This package is easier to use and is comparable to Microsoft Project. This program is a very nice fit for the individual project managers to use, while someone in the Project Management Office uses P3 to roll-up these files into an organizational program portfolio.

MANAGING TECHNOLOGY DEPENDENT OPERATIONS

Project KickStart – by Experience in Software, Inc.
www.projectkickstart.com

Project KickStart took 2nd place in Allpm.com's software survey "What project management tool/system do you find most useful?" Project KickStart was within 25 votes of taking first place from Microsoft Project! That says a lot about this intuitive and truly easy-to-use project defining and planning software. This is one of the few project management applications that walks the user through the defining and clarifying of the goal and scope. The dynamic links to fourteen other software packages means the user enters the data once and can use that data in many other applications (it even creates a project PowerPoint presentation for you!).

WBS Chart Pro – by Critical Tools
www.criticaltools.com

WBS Chart Pro is a Microsoft Project add-on software package that allows you to plan and display projects using a tree-style diagram known as a Work Breakdown Structure (WBS) Chart. WBS charts display the structure of a project showing how the project is broken down into summary and detail levels. Plan new projects using an intuitive "top-down" approach or display existing Microsoft Project plan files in an easy to understand diagram.

PERT Chart Expert – by Critical Tools
www.criticaltools.com

PERT Chart Expert is a Microsoft Project add-on software package that allows you to create presentation-quality PERT charts directly from your Microsoft Project plans. Loaded with features to configure and print many different styles of PERT charts (also known as Network Diagrams), PERT Chart Expert contains extensive PERT charting capabilities unlike those found built-in to Microsoft Project. If you plan on using PERT charts, this program is a must-have!

ManagePro – by Performance Solutions Technology, LLC
www.performancesolutionstech.com

ManagePro is an easy-to-use, versatile tool for managing projects, information, and people. This program helps the user carefully consider the goals and constraints of a project, as well as how the project fits in with the organization. The program creates the greatest value by organizing work processes around setting goals, working an action plan, and responding to feedback. Not quite a scheduling package like Microsoft Project, and not quite a fancy to-do list, this unique software aims to help an executive organize his or her thoughts.

VX-1 – by Virtual Experience Corporation
www.vx-1.com
This program is a computer-based project management simulation. This is the only project management simulation that allows (and requires) the user to plan their virtual project, then manage the plan, project team, boss, and customer. This program uses artificial intelligence and 108 patented "virtual people" to test, challenge, and hone the user's management skills. By completing this five-day simulation, the user has gained the equivalent of one year's project management experience. Using this program is the safest and most effective way to give existing or future project managers practical project experience, without fear of failure, and without detriment to the (real-world) organization.

GigaPlan – by Meridian Project Systems
www.gigaplan.com
This online collaboration service provides users with the ability to upload Microsoft Project (or compatible) files for the online viewing and updating of many other individuals. Rather than having to purchase additional copies of your scheduling software, those that need to view progress or even enter task updates can do so online through their preferred browser. Be sure to check out the free version of this service, which demonstrates how easy this process is.

Vertabase – by Standpipe Studios, L.L.C.
www.vertabase.com
This is a web-based project management application specifically designed to organize, streamline, track and centralize any type of project. This powerful tool improves team communication and adds visibility to all

steps of the production process. This application is a complete project planning and scheduling environment, including such features as interactive Gantt charts, personal daily task lists, resource allocation analysis, timesheet and vacation time management, budget analysis, automated email notifications, and flexible reports and charts.

Appendix 2
Quick Summary – Using Microsoft Project

As was mentioned in the previous Appendix, the authors formally recommend that the reader get directed training on project management software. The Newfoundland saying, "a fool with a tool is still a fool" certainly holds true. Project management software is neither intuitive nor forgiving, which is an especially troublesome combination. That said, this quick summary serves to guide the reader through the process of properly setting up a project schedule in Microsoft Project. These overall guidelines hold true for almost all other project scheduling software as well.

1. **Define the project** – As it has been mentioned earlier in this book, it is crucial to project success to clearly define the project goals and constraints in ways that are measurable and quantifiable. This necessary step must happen outside of Microsoft Project; there is no feature in the program that assists you in defining the project.

2. **Set the program options** – Go into the Tools menu and select Options. In this part of the program it is strongly recommended that you make the following adjustments:
    View – *show project summary task* should be selected
    General – *auto-add new resources* turned off, and at least $1/hour for a *Default standard rate*
    Schedule – *default task type* should be "fixed duration" and *new tasks are effort driven* should be turned off

3. **Set up the calendar** – Go into the Tools menu and select Change Working Time. This is where the overall daily start and finish times are set, as well as any holidays.

4. **Set the Project Information** – Go to the Project menu and select Project Information to enter the start date for the project. Be sure to keep the *schedule from* setting at "Project start date."

5. **Enter major deliverables in the task list** – These are also known as Level 1 tasks or summary tasks. An example would be "Foundation."

6. **Enter detail tasks in the task list** – These are also known as Level 2 tasks or activities. An example would be "Dig hole." Be sure to indent the activities (Alt-Shift-Right Arrow) to indicate which major deliverable they are a part of.

7. **Reorder the tasks** – Rearrange the order of the task list, to put the tasks that happen earlier in the project (chronologically) towards the top. Of course, maintain the major deliverable (summary) groupings.

8. **Enter durations** – In the duration column, enter the estimated duration of the detail (Level 2) activities. Do not type in any dates!

9. **Set dependencies** – Using the Linking Tool, indicate which tasks must finish before another task can start. This can also involve fine-tuning the type of dependency by going into the Task Information dialog box for the dependant task.

10. **Define resources** – Going to the Resource Sheet, type in the names and details for each of the resources.

11. **Assign resources** – In the Gantt chart view, select the desired task, then assign the appropriate resource(s).

12. **Refine resources** – In the Gantt chart view, with perhaps using additional split-screen views, look for over-assignments and make the appropriate adjustments.

13. **Examine the Critical Path** – Using any of the appropriate built-in features (filters, highlights, formatting, columns), determine which tasks are on the Critical Path and which have float or slack in them. Check the amount of float or slack.

14. **Optimize, add flexibility, verify feasibility** – Review the overall schedule and look for ways to optimize (shorten the project duration, decrease the amount of money needed, decrease the number of resources needed, etc.) while adding back in flexibility (adjusting durations and dependencies to ensure the inevitable slippages will not cause a domino-effect of date shifting). Verifying feasibility is the user's last chance to adjust and fine-tune the project schedule so that he or she is confident that this schedule is realistic.

15. **Set baseline** – Use the Tools menu, Tracking option, then select Save Baseline. The schedule is planned, and any adjustments made to the file will be interpreted by the program as the entering of "actuals."

Please keep in mind that this very brief checklist is intended to remind the reader of what should be done in what order. This checklist is certainly not a substitute for classroom training on the software, or a self-learning how-to book such as Hungry Minds *Teach Yourself Microsoft Project* written by Vickey Quinn.

Appendix 3
Project Management Online Resources

The following online resources come highly recommended. Whether you are relatively new to project management or have been working on projects for many twenty years, these websites contain techniques, tips, and best practices that almost anyone would find worthwhile.

Please be aware that because of the constantly changing nature of the Internet, some of these websites might not always be available. Consult your favorite search engine to help track down a "missing" site listed here.

www.4pm.com/repository.htm
This project management website is chock full of best practices, templates, examples, and worthwhile articles.

www.allpm.com
Originally started years ago as a way for Michael Lines to personally support and expand the profession of project management, the site has since grown into a living collection of project management and quality management resources, discussions, templates, and even job listings.

www.appl.nasa.gov
This is the home of the NASA Academy of Program and Project Leadership. This U.S. government organization has posted several "Knowledge Sharing" resources, all furthering project management best practices.

www.acq.osd.mil/pm/
This website provides information on earned value project management for government, industry and academic users. Be sure to click on the heading "What is Earned Value Management?" right under the main title for another good explanation of this project tracking and managing technique.

MANAGING TECHNOLOGY
DEPENDENT OPERATIONS

www.pmforum.org
The PMFORUM is a resource for information on international project management affairs. The PMFORUM supports the development, international cooperation, promotion, and advancement of a professional and worldwide project management discipline.

www.projectconnections.com
This for-pay website is both a clearing house of templates and sample project documents, as well as an online discussion forum. The many free "teaser" downloads makes this website a must-visit destination.

www.projectmanagement.com
With several commercial ties, this website does still have some useful information and templates that can be downloaded at no cost.

www.project-manager.com
After clicking on the "map" comment, the visitor is taken to a flowchart navigation graphic, which makes finding useful information a breeze. (For ease of navigation on this site, use the "back" feature of your browser to back out to the main navigation graphic.)

www.uscg.mil/hq/g-a/Deepwater/Welcome.htm
This website holds the complete project documentation of the U.S. Coast Guard's Deepwater project. If the reader wishes to see the project planning for a massive multi-year program, download these documents.

www.jsc.nasa.gov/bu2/guidelines.html
NASA's Johnson Space Center has put together this website, which holds many thorough explanations and techniques for cost estimating. Be sure to explore their list of additional cost estimating resources.

www.wa.gov/DIS/PMM/index.html
This website, created and maintained by the State of Washington, features how-tos and best practices for portfolio management. It also holds

several resources for the Software Engineering Institute's (SEI) capability maturity model (CMM) as it relates to software projects.

www.ist.uwaterloo.ca/projects/templates.html
This website holds project templates for Information Systems and Technology projects. Maintained and hosted by the University of Waterloo in Waterloo, Ontario (Canada), these free templates are available in Microsoft Word file format, as well as web-page HTML.

Appendix 4
Project Management-related Professional Societies

The following professional societies are related to the furthering of project management best practices. The reader should considering joining and becoming active in one or more of these organizations to better develop knowledge and techniques for effective management.

Project Management Institute – PMI®
Four Campus Boulevard
Newtown Square, PA 19073-3299
610-356-4600
www.pmi.org
This international organization has been around for over thirty years. They have collected best practices, conducted research with universities, and have furthered the standardization of project management practices around the world.

MANAGING TECHNOLOGY
DEPENDENT OPERATIONS

International Project Management Association – IPMA
PO Box 1167
3860 BD NIJKERK
The Netherlands
+31 33 247 34 30
www.ipma.ch

    The International Project Management Association (IPMA) is a non-profit, Swiss registered organization, with a Secretarial office based in the Netherlands. Its function is to be the prime promoter of project management internationally through its membership network of national project management associations around the world. Additionally, it has many individual members, people and companies, as well as co-operative agreements with related organizations world-wide, to give it a truly world-wide influence.

International Society for Performance Improvement – ISPI
1400 Spring Street, Suite 260
Silver Spring, MD 20910
301-587-8570
www.ispi.org

    Founded in 1962, the International Society for Performance Improvement (ISPI) is the leading international association dedicated to improving productivity and performance in the workplace. ISPI represents more than 10,000 international and chapter members throughout the United States, Canada, and 40 other countries. ISPI's mission is to develop and recognize the proficiency of our members and advocate the use of Human Performance Technology. Assembling an Annual Conference & Expo and other educational events, like the Institute, publishing books and periodicals, and supporting research are some of the ways ISPI works toward achieving this mission.

PROJECT CONTROL
By Eric A. Spanitz

International Society of Parametric Analysts
636-527-2955
www.ispa-cost.org

The purpose of ISPA is to improve our effectiveness and leverage the limited resources of our respective organizations by providing information on techniques, tools, and methodologies for enhancing the management decision-making process.

Society of Cost Estimating and Analysis – SCEA
101 South Whiting Street – Suite 201
Alexandria, VA 22304
703-751-8069
http://www.sceaonline.net/

This is a non-profit organization dedicated to improving cost estimating and analysis in government and industry and enhancing the professional competence and achievements of its members. This organization complements the PMI® by providing scientific research in the realm of estimating and cost analysis. Their occasionally published "Journal of Cost Analysis and Management" contains papers and articles about the latest advances in estimating, usually with a mathematical foundation. The "National Estimator" is published a little more frequently, with more everyday articles about estimating, cost analysis and earned value.

Society for Risk Analysis
1313 Dolley Madison Blvd. - Suite 402
McLean, VA 22101
703-790-1745
www.sra.org

The Society for Risk Analysis (SRA) provides an open forum for all those who are interested in risk analysis. Risk analysis is broadly defined to include risk assessment, risk characterization, risk communication, risk management, and policy relating to risk.

MANAGING TECHNOLOGY
DEPENDENT OPERATIONS

American Society of Professional Estimators
11141 Georgia Avenue - Suite #412
Wheaton, Maryland 20902
301-929-8848
www.aspenational.com

    The focus of the American Society of Professional Estimators is to serve construction estimators by providing education, fellowship, and opportunity for professional development. Although their main intent is to serve construction estimators, their best practices for estimating are applicable for all types of projects.

Chapter 12, Lead-in:

LOGISTICS AND THE INTEGRATED ENTERPRISE

It hardly is news that the integration of systems is dependent on computers and information technology. For operations the systems advocates place their bets on the Enterprise Resource Planning (ERP) tools. The major consulting firms have built an industry assisting clients in the application of systems tools. You must know how to approach them and evaluate their contribution. How do you plan to keep up when the consultants leave?

Information technology in the form of cookie-cutter software packages and the initial push by outside consultants is not enough. To manage operations systems takes a blend of skills. It has to be linked with an appreciation of the processes, materials, and hardware involved. In the modern, slimed-down manufacturing operation over 50% of the cost of goods sold will be in the category of materials. Outsourcing needs to be managed. For cost reduction and performance improvement the general manager in this chapter is shown range a of technologies and methodologies to manage the supply chain.

This chapter points out that, while the Internet has effectively reduced the information transaction cost to nearly nothing, the demands on a manufacturing or service enterprise requires that a full toolbox approach be taken. Today's executive must realize that no single tool, whether a technology or methodology, is the "silver bullet." There are those advocating lean methodologies, with visual control, kanban cards, JIT layouts and one-piece flow. The parochial systems advocates have no room or interest in lean methodologies. The JIT folks have no interest or room for the systems solution. The true answer to optimal performance is a full toolbox and an appreciation of when and where to use them.

MANAGING TECHNOLOGY
DEPENDENT OPERATIONS

Chapter 12

LOGISTICS AND THE INTEGRATED ENTERPRISE

By Richard Hammond

We are in a new business environment as a result of technology-driven labor cost reduction, the application of lean manufacturing methodologies, value engineering efforts to reduce piece parts, and off-shore sourcing of high labor content products. Materials and supply chain costs now account for well over 50% of the cost of goods sold. The labor content of cost of goods sold is well within a single digit percentage. The recent operating focus is on –

- material costs,
- associated supply chain costs, and
- internal and external quality costs.

The following points attention to material costs and associated supply chain activities.

I – The New Environment

In addition, marketing efforts within our companies are segmenting markets. Each segmented market is being targeted with specific products, packaging, and services. Each channel of distribution needs to appear to be uniquely designed for that customer or set of customers. Even though

LOGISTICS AND THE INTEGRATED ENTERPRISE
By Richard Hammond

the customer interface may be structured for that particular channel, the systems and processes behind the scene should remain standard and disciplined to base company policies. The customers are becoming even more demanding.

> *Materials and supply chain costs ... over 50% of the cost of goods sold.*

The customer base is asking for even more complex product offerings, as our base company attempts to standardize our product. Not only are customers looking for a differentiated, high function product; but the customer or consumer is looking for the lowest price. There is a continuing pressure on price, associated margins and cost of goods sold. As outlined in the first paragraph the labor costs have generally been wrung out of the product. What remains as a target is material and supply chain costs. So not only are customers asking for a high service level, but it is these supply chain features that are being targeted for cost reduction, rationalization, and elimination. If the scenario is not complex enough, the supply base is also becoming more complicated.

In one aspect the supply base is becoming rationalized to a smaller number of vendors. Dr. Deming advocated single supply sources and the partnering aspect with the supply base. As we have narrowed the supply base for our companies, we now rely on that critical single source. Communications with that supplier become more critical as we exchange information, forecasts, data, orders, and releases on a nearly real-time basis. Because the supply base has been reduced and a single vendor gets the leverage for commodity buying across the division or corporation, we usually achieve an associated significant cost reduction for our actions. Supplier choice and partnership is critical as we move to single sources for a particular part, rather than across a commodity base. We must monitor the vendors' business health, as well as, his lead time, cost, delivery performance, and quality. Even though we likely achieved a significant cost reduction as a result of rationalizing the supply base, the continued pressure on material cost reduction remains.

MANAGING TECHNOLOGY
DEPENDENT OPERATIONS

The next step in developing material cost reductions is to begin looking at offshore or non-U.S. suppliers or assembly shops. We hear many stories of U.S. manufacturers and suppliers locating operations in Mexico. Nations in eastern Europe, notably Poland and Romania, have begun offering parts and sub-assemblies at reduced costs and improved quality. Along the Pacific Rim, with the slow down of the world economy, specifically in electronic technology, China and South Korea are becoming even larger sources of printed circuit boards, electronic modules and components. China has not limited themselves to the electronics and appliance industries; they are a formidable source for machined castings and forgings. Romania has also developed export capabilities in processed metals. Even though basic processes may be backward and crude to western standards, high quality metallurgical properties are being achieved at reduced costs.

> *Supplier choice and partnership is critical...*

So the scenario being developed is challenging. Customers are more demanding, the supply chain is now strung out across multiple continents, and our corporate leadership continues to press for lower costs, higher margins, and increased performance to win over the customer base. The environment is even more complex. Corporations, as a result of poor business ethics and performance since the bubble burst in 2000-2001, are under greater scrutiny. Cash flow has become a critical aspect of business performance. The two aspects of cash flow focus on the supply chain, being total inventory and accounts receivables. If the supply chain works and the customer is happy, they will generally pay according to terms, and the accounts receivable level will hit a reasonable steady state. If the supply chain or product fails, problems will cause the accounts receivable levels to rise. Inventory has more direct bearing on cash flow. Inventory may

> *If the supply chain works ... accounts receivable level will hit steady state.*

still be an asset on the balance sheet. But inventory in excess of requirements is an undesirable factor in business management.

High inventory levels, with low turnover, results in low inventory turns. Excess inventory is not an asset in operating the business. So as the supply chain is stretched from overseas to a more demanding customer, operations management must also face the daunting task of reducing the inventory level of the enterprise. This inventory will generally consist of raw material, parts, work in process, finished goods, and non - production inventory. The short answer is to press our supplier and customer to hold more inventory. But our vendor and customer are faced with the same business pressures we have. There is always a cost to holding inventory. The solution is to reduce the inventory level throughout the complete supply chain. But how is this accomplished? There are many tools available to get this task complete. If we borrow from the works of Eliyahu Goldratt and his book, *The Goal*, what we will try to do in our enterprise is optimize through-put while reducing inventory and expenses. Lets begin looking at the choice of tools to accomplish this task.

The challenges are daunting, but the tools are plentiful. As a business leader you need to assure that the project manager assigned the task of optimizing the supply chain has training in a range of technological tools and facilitation methodologies.

> *... pull in ideas from a multifunctional team...*

He or she needs to draw on lean methodologies to pull in ideas from a multifunctional team utilizing team facilitation skills. He or she needs to understand the latest capabilities and requirements of manufacturing or enterprise systems (MRP/ERP). He or she needs to have a feel for bar code technologies, proximity sensing devices, geo-positioning and satellite communication systems. There is a policy, procedure, and discipline factor required to achieve optimum results. People must interact with any system developed, so we again see a need for facilitation, management, and leadership skills.

MANAGING TECHNOLOGY
DEPENDENT OPERATIONS

The program leader must possess a full toolbox of skills and an understanding of technology and lean manufacturing methodologies.

No one tool can do it all. Do not be sold on a one dimensional solution. See Table 1.

Table 1: Examining The New Environment

1. What is the profile of cost of goods sold for your product or a typical product in your offerings? Has this profile changed over recent years?

    No ___; Yes ___; How? _____

2. What is the breakdown of current levels on the inventory components; raw material, work in process and finished goods? How many inventory turns does your enterprise produce in a typical year? Do you know?

    Yes ___; No ___; Who knows? _____

3. What are the trends in absolute inventory levels and inventory turns for your business?

    Have ___; To get ___; Where? _____

4. Is your business simplifying the vendor base by reducing suppliers, leveraging your buys with sole sourcing, and developing partnerships?

    Yes ___; No ___; Where to start? _____

5. Do you know the business requirements of your customers? Is your enterprise creative in the use of people and systems to create a liaison operation to meet your customers' needs?

    Yes ___; No ___; Who can help? _____

LOGISTICS AND THE INTEGRATED ENTERPRISE
By Richard Hammond

II- Full Tool Box Approach

The individual assigned to optimize the supply chain or the logistics operation needs to be equipped with a full toolbox. Not only must they understand the material resource planning (MRP) or enterprise resource planning (ERP) software and integrated system, but they should be comfortable with the lean manufacturing methodologies. They should know network/internet communications, data base management principles, sales and operations planning, and an understanding of bar code/RF data collection and communications technologies. The following discussion will outline a typical approach to achieve systems optimization in the current industrial environment to illustrate how each of these methodologies and technologies are utilized. No single narrow approach will achieve full optimization of the supply chain, in spite of product or consultant claims.

First:

The organization must understand the supply chain in conceptual terms. One available tool to achieve this knowledge is the utilization of Value Stream Mapping. A valuable reference to this technique is titled *Learning To See – Value Stream Mapping to Add Value and Eliminate Muda* by Mike Rother and John Shook. By the way, the simple definition of "muda" is operational waste.

> *... understand the supply chain .... available tool is Value Stream Mapping...*

The procedure is to gather a multi-discipline team including sales/marketing, manufacturing, and purchasing. The exercise is to map the flow of the parts from the supplier to our dock, a macro mapping of the manufacturing process and then the flow of the finished goods to the customer. Supplier data such as lead-time, shipment size, travel time, etc. is collected to gain understanding of the supplier base. Critical suppliers are mapped for a firm's product to understand the responsiveness of one portion of the supply chain. The value stream map captures critical data relating to the customer base and defined

MANAGING TECHNOLOGY DEPENDENT OPERATIONS

requirements. Considerable work goes into really understanding the customer in each distribution channel. Critical manufacturing data is gathered to understand the objective performance of the operation. Across the top of our value stream map, we illustrate the current communication scheme to identify how the customer communicates his needs to our company, and how we communicate our requirements to the supply base. How are actionable instructions given to the manufacturing plant and transportation entities to provide material movement in the process? Facilitation skills are essential in this process. This team needs to meet on several occasions to assure a complete, comprehensive picture is being drawn of the value stream. It is critical to understand all aspects of the customer. Choose one typical representative product family as a subject of the map. The map will take up a full wall in a conference room setting. Use "post-it notes" to capture system elements and operational data.

Second:
Critical operational components need to understand the current and future needs of the customer. The process utilized is the Sales and Operation Planning meeting. The meeting consists of key personnel from manufacturing, sales, finance, and materials. The current customer requirements are initially discussed. Manufacturing and materials needs to commit to the short-term plan.

> *It is critical to understand all aspects of the customer.*

If there is a problem, expediting requirements must be identified or causals for shortfalls discussed with sales. The finance representative provides an assessment of whether current commitments and capabilities achieve budgeted revenues and margins. Everyone understands the requirement, expectations and possible shortfalls. This meeting provides the forum to understand the issues and challenges facing the organization. The meeting development begins with a short-term focus and quickly begins to review planning at successively longer horizons. The best I have seen provide a detailed review of the current quarter plus four additional calendar quarters. The greatest challenge to the process is the participation of the sales representative and the value of the sales forecast. This forecast must truly reflect reality, not a stretch goal for the organization. This is the forecast which will be loaded into the MRP/ERP system to communicate with the supply base and begin the flow of materials to the plant. No matter how advanced your MRP/ERP system has been sold to the organization, the face-to-face sales and operations planning meeting is critical to the success of the logistics portion of the enterprise. The meeting needs to begin as a weekly event. As the team develops, the meeting moves to a monthly schedule. The rough cut capacity report off the MRP/ERP system will eventually be a critical tool utilized by this team. The developed sales and operations process with a one-year horizon becomes the input for next year's budget and operations plan. The process also provides the input to capital budgeting requirements. For a better understanding of the sales and operations planning process, a great reference is *The Oliver Wight ABC Checklist for Operational Excellence.*

> *forecast must truly reflect reality... loaded into the MRP/ERP system ...to begin the flow of materials ...*

Third:

The software selection process.
Now that there is an understanding of the value stream from suppliers, through manufacturing to the customer; software selection can take place if the system is not already established. In the software selection process one immediately has SAP, People Soft and Oracle come to mind. There are many other smaller software firms which may be a great fit for your enterprise. The "APICS, The Performance Advantage" monthly periodical dated June 2003 contains a buyers guide for MRP and ERP systems. The organization's web address is: www.apics.org if you want to receive an excerpt of the buyers guide. The organization must understand their value stream and have a strong sense

> *Systems/software selection is a daunting task... SAP, People Soft and Oracle come to mind...*

MANAGING TECHNOLOGY DEPENDENT OPERATIONS

of their systems requirements, before the software sales representatives calls on your enterprise. System/software selection is a complex task beyond the scope of this chapter. It is critical to understand that the system/software is only one tool in the toolbox. There are many procedures and policies that interface with the system/software including suppliers, customers, people, and machinery. The selection of the software and required modules should also take place in a team environment with a strong project leader. The organization needs to determine the degree of system integration required, the modules selected to provide a set of business functions, and the project plan for implementation. Recent systems integration projects are led by <u>operations personnel</u>, with IS(information systems) personnel serving in a technical support role.

Fourth:

The Internet has provided the key tool. It can truly integrate the supply chain from supplier, through manufacturing to the customer. The Internet provides very low cost, efficient information and data transmission between supply chain components. As a business leader it is important to understand that the trusted relationships are as important as the capabilities of the technology.

> *...study the management process of Wal-Mart and read Michael Dell's book titled <u>Direct from Dell</u>...*

The ideal would see a customer sell a product. That transaction initiates a series of messages to replenish the inventory. Internet messages would flow directly to the supply base to initiate the flow of parts, components, and raw material to the manufacturer. A parallel message would be sent to the manufacturer to schedule the item into the production schedule. Messages would also be directed to the transportation providers to begin developing loads and destinations. Pockets of industry are striving towards this transparent, fully integrated system. Business leaders need to study the management process of Wal-Mart and read Michael Dell's book titled *Direct from Dell* for a deeper understanding of these processes. It takes many

LOGISTICS AND THE INTEGRATED ENTERPRISE
By Richard Hammond

years to achieve the value stream or supply chain efficiencies of a Wal-Mart or Dell Computer. Your enterprise needs to get started and to establish a continuous improvement culture focused on material logistics costs. American industry has focused on labor cost reduction for many years bringing that factor to single digit percentages of the cost of goods sold. We need that same focus on material and transportation costs. For a better understanding of continuous improvement efforts and essential book to review is *Out of the Crisis* by Dr. William Deming.

Fifth:
There are many technologies to support the efficiency of the supply chain. This will not be a complete review of those items but several notable items will be presented. Bar code technology will increase the accuracy of data collection and speed up the process at receiving/shipping docks. The bar code reading devices are also used to support verification of inventory inside our operation or to trace material flow between major portions of the manufacturing operation. Bar code technology has became so reliable and user friendly that the devices are used in nearly all grocery and department store check out areas. The devices are hard wired into an interface device or RF (radio frequency) transmission is utilized for portable readers. The use of bar code reading for material tracking raises the same type of question raised when discussing internet information flow: how much data becomes too much and overwhelming versus what is needed to efficiently run the operation.

> *...how much data becomes too much?*

The selection of transportation carriers finds us asking similar questions. Most consumers have utilized UPS and Federal Express. An item is dropped off or picked up by the carrier. A Pro # is assigned to the package. You, as the customer, either by phone or personal computer can track your shipment and estimated delivery time. Carrier choice is also critical. What level of shipment visibility and tracking is required? Carriers equipped with Qualcomm technology devices linked to satellites are capable of providing real time shipment tracking. This service comes at a cost. Is this information required to run the enterprise? Is there information needed to place the company on a path of continuous improvement? Whatever devices are utilized to collect data and support the system need to be interfaced electronically. Competent information services support is required to interface devices from two separate vendors. This task is never easy, in spite of the claims of the local sales force.

> *place the company on a path of continuous improvement...*

The concluding thought is to understand your supply chain, utilizing the value stream map, and understand the needs of your business before making any solicitation of technologies devices. (See Table 2). Real time visibility of orders and materials may only be necessary for a portion of your supply chain. Choose your tools wisely. They come at a cost in both price and the effort requirement to install and interface to the complete system.

Table 2: Supply Chain Management Review

1. Does _____ have the breadth/depth to take on the supply chain integration project?
    Yes __; No __; The option:_____
2. Do you have an I.S. staff capable of supporting a supply chain integration project?
    Yes __; No __; The option:_____
3. Can you identify three key suppliers, their lead times for raw material or components?
    Yes __; No __; The option:_____
4. What are the current and future channels of distribution? Are these well managed?
    Yes __; No __; The option:_____
5. What are the customer demands for each of these channels integrated?
    Yes __; No __; The option:_____
6. Does your enterprise have a continuous improvement culture?

    Yes __; No __; The option:_____

LOGISTICS AND THE INTEGRATED ENTERPRISE
By Richard Hammond

III - Policies, Procedures and Disciplines

As stated in the first two sections, the logistics/supply chain leader must possess a full toolbox of technological items and lean methodologies. The technological items would include the systems software, computer hardware, communication protocols (internet) and automatic data collection (bar code). The methodologies would include leadership vision of how the supply chain would be structured in the future to support the business. Another key aspect of an efficient supply chain are the policies, procedures, and disciplines required in running the system. A comprehensive review of these items is contained in *The Oliver Wight ABCD Checklist for Operational Excellence*. The book includes the people and communications processes required to support an integrated system centered on an MRP/ERP system. The book is very comprehensive in defining a class A operation. This section will outline some key priorities to achieve a high level utilization of the system and to attain supply chain excellence. The items noted will have the highest impact to the operation, warranting the attention of the manufacturing executive.

1. The first item is the sales and operations planning meeting that sets the direction for the organization and assures we are all working off the same plan. This was discussed in greater detail in section 2 of this chapter.

2. The second critical factor is bill of material accuracy. When forecasted demand for a finished good is placed in the system as a projected build, a "parts explosion" takes place. The list of parts for the finished good is pulled from the finished good bill of material. The

> *If there is a systemic problem with bills of materials, inventory may continually rise...*

parts are organized/collated within the system and suppliers are notified of our forecast and/or items are placed on order. If the BOM (bill of material) is inaccurate in any way the wrong parts are ordered

MANAGING TECHNOLOGY
DEPENDENT OPERATIONS

or parts needed are not ordered at all. When the internal requirements indicate that the operation is to produce or assemble the finished good, if the BOM is inaccurate, the assembly task is impossible. One of two scenarios takes place - both of which result in the finished good not being produced on time. One possibility is that a part is not available. The organization then goes into a fire fighting, expediting mode to acquire all required parts. Another possibility is that the wrong part is procured, with the same result as the first example. The assembly cannot be produced and the organization needs to expedite the correct part. The part received becomes part of inventory. If someone does not take action to have the part returned, the item stays in inventory. If there is a systemic problem with bills of materials, inventory may continually rise at the same time that the correct inventory is not available to produce finished goods. To meet customer schedules, and to reduce inventory bills of material must be accurate. The standard for a world- class operation is 98-100%. The measurement is tough and the recognized calculation is:

$$\text{BOM Accuracy \%} = \frac{\text{Number of Correct Bills}}{\text{Total Number of Bills Checked}}$$

"Number of correct bills" implies 100% perfect accuracy in the bill of material and that comparing the bill of material in the system against the latest engineering documentation for that product.

> *...monthly readings will provide a feel of whether BOM accuracy is improving...*

Bill of material auditing is an ongoing endeavor. Bills are audited with major corrections taking place within the engineer change notice (ECN) process for major discrepancies or corrected by the auditor for routine items. A running tally is taken, and generally a BOM accuracy reading

LOGISTICS AND THE INTEGRATED ENTERPRISE
By Richard Hammond

is calculated for the month. A series of these monthly readings will provide a feel of whether BOM accuracy is improving or getting worse. An organization can spend an inordinate amount of money on a MRP/ERP system. The system will never yield the results promised if BOM accuracy is suspect.

3. The third critical factor is inventory accuracy and cycle counting. The logic surrounding inventory accuracy parallels that of BOM accuracy. A core group of individuals in your materials organization are assigned cycle counting. A system generated list of part numbers and components are identified each day. The personnel physically go to the inventory location designated by the system and count the material. If the item, quantity, and location are accurate the organization receives one successful entry. If any of the three factors are incorrect, no credit is given. The standard calculation is:

$$\text{Inventory record Accuracy} = \frac{\text{Number of correct items/quantities/locations}}{\text{Number of items/quantities/locations checked}}$$

If there is a schedule requirement for a part, and the part cannot be found the organization must search for the part and/or go into an expedite mode. Finished goods are late in shipping to the customers. Expedited freight costs are realized, which reduces the overall business margins. Inventory accuracy, much like BOM accuracy, is imperative to running a truly efficient operation. The software vendor will sell the integration and sophistication of the MRP/ERP system. The software house will sell your organization on the reduced headcount your enterprise will achieve to run the operation. As the operations leader you need to keep a core group of people to audit and lead the correction of bills of material and inventory accuracy.

4. The fourth item to consider is key factors measuring through-put time in manufacturing routers and actual lead times in the item master. These two factors are absolutely critical for short-term production scheduling and sequencing, and longer term "rough cut capacity planning." During the introduction of a new part the item master will

MANAGING TECHNOLOGY
DEPENDENT OPERATIONS

> *item master and routers are another set of items which must be maintained*

be filled in with all current data associated with the part. In the same matter, the industrial or manufacturing engineer will fill in all the fields in the manufacturing router, including process/throughput times. These factors change, so the item master and routers are another set of items which must be maintained. If erroneous data is being utilized short-term shop schedules will be in error and capacity-planning tools generally utilized for mid to long term planning will have no value.

5. The fifth critical set of items falls in the category of policies or basic shop practices. Systems integration people boast of the visibility that their systems provide. Unfortunately it takes hard work to provide that accurate picture of the operation. There are three basic items to focus your attention on. They seem very simple, but they will make a difference on the quality of your systems performance. The first factor in this category is to clear the receiving dock each day, log the receipts into the system and accurately put the parts away. When MRP is run each day or each week, there will be a demand for parts. If the parts are in the system, production is scheduled and the parts are allocated. If the parts are not accounted for in the system, the parts are placed on order. If the parts are physically on the dock, but not accounted for in the system, the parts will be placed on order. The order is cut today, the parts already on the dock are put away tomorrow, in several weeks you experience excess inventory for that part. Multiply this condition by all the parts on the receiving dock and you soon have excess inventory problems. Following the same logic as the parts received, the finished goods need to be reported and parts "back flushed" or relieved from inventory. If completed finished goods are not reported and part counts reduced

> *In lean (JIT) manufacturing daily scrap reporting may not be frequent enough.*

LOGISTICS AND THE INTEGRATED ENTERPRISE
By Richard Hammond

the system still sees these parts in inventory and assumes they are available for the next finished goods order. The third factor in this category includes the timely reporting of scrap. If a part is scrapped and not available for production, the system sees the part as available for use until it is reported as scrap. In the old days of manufacturing legacy system, reconciliation and reporting of scrap once each week was adequate. In the current lean manufacturing, JIT environment daily scrap reconciliation and reporting may not be frequent enough.

By understanding how some basic operational imperatives and data integrity can effect your manufacturing system, you will come a long way in your ability to challenge the operation for excellence and continuous improvements. Eventually one hopes to move your MRP/ERP system from a basic transaction tool and short-term management device to order parts and schedule the shop, to a strategic tool to assist with planning, plant simulation, and eventually, plant optimization. Consider examining the status of your materials monitoring discipline according to Table 3.

Table 3: Monitoring Discipline

2. What is your operation's BOM accuracy?
   Know___; Don't know___; Who can tell?_____

3. What is your operation's Inventory accuracy?
   Know___; Don't know___; Who can tell?_____

4. Who is auditing and maintaining your operations BOMs and inventory?
   Know___; Don't know___; Who can tell?_____

5. Do your receiving personnel stay at work each day until the receiving dock is clear and all supplier items have been received in the system?
   Know___; Don't know___; Who can tell?_____

6. Does scrap parts and material accumulate in the plant?
   Know___; Don't know___; Who can tell?_____

7. Does rework accumulate for long periods?
   Know___; Don't know___; Who can tell?_____

8. Does rework move back into the mainstream of work in process in a timely matter?
   Know___; Don't know___; Who can tell?_____

MANAGING TECHNOLOGY
DEPENDENT OPERATIONS

IV – Supply Chain Examples in our Daily Lives

In sections one through three we discussed the new economic and specifically manufacturing environment which places greater focus on material cost. This very focus on the material aspect of our enterprise leads us to improve the efficiency and effectiveness of the supply chain, section 2. Then section 3 discussed basic operational imperatives and disciplines to optimize the technology and improvements which have been put in place. Unlike manufacturing technologies, which a small segment of the population understand, we come in contact with examples of great supply chain practices in our daily lives.

Bar code technology is used in nearly all retail establishments. Not only is the transaction speed optimized, but also the data collection provides a history of demand for future marketing or sales programs. Inventories are monitored on a nearly real time basis, and stock is replenished without manual intervention. Another inventory replenishment technique I still see in some small convenience grocery/snack stores are Kan-Ban cards. Packages of peanuts and candy are hung for display. A small card, sometimes displaying a bar code is positioned at the reorder point for that item. Periodically the store attendant collects the exposed Kan-Ban cards and inputs the bar code into the management system to trigger the reorder cycle.

Another exciting technology involves placing tiny radio frequency identification tags on items. Readers or input devices tied to central systems monitor the movement of inventory through the shop or supply chain. During the Chicago Marathon, each runner is given an identification tag which they attach to their shoe. A runner's time is accurately determined by "reading the tag" when the runner passes the starting line, and taking the last reading at the finish line. Race results for the casual runner are available by the time the runner finishes, returns home, and boots up their personal computer. To assure none of the runners take a short cut across the Chicago Loop, additional readers are placed along the marathon route to monitor each runner. Think of the other possibilities that this identification system provides. If a runner has an injury or accident, emergency personnel and race

LOGISTICS AND THE INTEGRATED ENTERPRISE
By Richard Hammond

administrators have instant access to key data on name, age, etc., but also phone numbers and points contact from the registration data base. Toll road systems are now using identification technologies for frequent users of the toll road network in Houston and Chicago. By registering your automobile and developing an automated (credit card) payment scheme, you receive an identification tag that hangs from your rear view mirror or is attached to your front bumper. Instead of sitting in line to drop coins into a chute, you pass the tollbooths at a reasonable speed for the reading device to identify your automobile and debit your credit card.

Just imagine how these identification techniques that have proven reliable in the fast pace retail and personal environment could be used to monitor your raw work in process and finished goods inventory.

Another system you may want to test is to ship a package by UPS or Federal Express. Drop off the item, and you receive all appropriate documentation at the time. Then begin tracking your shipment through your personal computer and the shipper's web site. The small package shipping business has been applying the latest advances in identification techniques, readers to trace movement, radio frequency operations, data base management, and in the end - a real life person to interact with the customer.

Purchase a personal computer from Dell or Gateway and you experience two companies that have optimized their supply chain and utilize a combination of technology and personal phone contact to provide superior customer service and satisfaction. Dell and Gateway provide a limited offering of products with standard accessories. Their vendors keep the supply chain full and optimized to assure a quick turnaround when you order on-line or via phone.

One's life experiences will provide multiple examples of outstanding, cutting edge supply chain management as one experiences personal business transactions with Dell Computer, Gateway, Home Depot, Wal-Mart, UPS, Federal Express, or your corner grocery store. These firms are using proven technologies at a mature cost low enough to be utilized in the retail

MANAGING TECHNOLOGY
DEPENDENT OPERATIONS

market. It is up to us as business leaders to determine which methodologies and technologies can enable our enterprise to provide superior customer service and to optimize the supply chain for greater margins.

References and Suggested Reading

1. Goldratt, Eliyahu M. *The Goal.* North River Press, 1992. Second Revised Edition.

2. The Oliver Wight *ABCD Chedcklist for Operational Excellence.* John Wiley & Sons, Inc. Fifth Edition, 2000.

3. Dell, Michael. Direct from Dell, Strategies That Revolutionize an Industry. *Harward Business Review*, 1999.

4. Yue, Iaroene. *Tags Pit Efficiency vs. Privacy.* Chicago Tribune. Business Section, July 15, 2003.

5. *APIC, The Performance Advantage.* www.apics.org.

6. IE, Industrial Engineer. www.iienet.org.

EPILOGUE

SO WHAT DOES THE FUTURE HOLD?

The rate of technology change has been incredibly rapid over the past few decades, and there is every indication that change is accelerating. This point has frequently been made and there is no reason to question that this is the future. This relatively obvious observation was the reason for preparing this book, since traditional management approaches have been based primarily on steady state operations

In the previous chapters of this book, the editors have put together a collection of "tools" that may prove useful for managers of technology dependent operations. This book is clearly only a toolbox – the selection and application of the appropriate tools is left to the reader. This book is not and cannot be the definite "cookbook" for management of technology dependent operations. A careful reader will, in fact, find significant difference between the contributors to this book - there are no "school solutions" and the reader is at best given some ideas that may help him or her as they attempted to solve their management issues. This book has given the reader some understanding of the issues and a collection of tools that might be useful – the reader is left on his or her own to determine how to best use these tools to build their organization.

As was pointed out earlier, all operations have become or are becoming technology dependent. Technology dependency and change are rapidly becoming the norm. This observation seems self-evident to the editors

MANAGING TECHNOLOGY DEPENDENT OPERATIONS

and convinced them and the other contributors to offer this toolbox. With very rare exceptions this "obvious" point is not covered in traditional management publications or research. The "norm" in most management publications is still based on steady-state operation where change is considered as a special situation and where technology is treated as a something requiring special controls.

It is the editor's view that there has to be greater emphasis on management education and research based on this new paradigm, (i.e. the combination of new technology and change being the normal conditions under which management has to operate). The contributors to this book have raised some of the issues that should be studied in greater detail. As a summary of these thoughts and as suggestion for further analysis the following topics are suggested:

What is Technology?

The almost automatically "jump in logic" made when the word technology is used is to "information technology." The second "jump" is to "high technology." For the purposes of this book, "technology" is used in a much more pedestrian way to mean the application of any knowledge into an organization. Based on this understanding, processes that have long been commercialized, an organization can introduce "new" technology that other organization consider "old hat."

What is the Relationship Between Technology and Change?

Introduction of a new technology always requires a change in the status quo. Change is always uncomfortable – "old habits die hard." It is important to note that change can be equally or even more difficult for managers. In particular, managers who have succeeded in traditional organizations will be challenged.

EPILOGUE

How has Technology Change been Handled?

In the world of traditional management, new technology was introduced in a logical fashion. Management decided what technology was appropriate and took actions to introduce that change. Those down the "chain of command" were involved to implement a decision that was already made. There was discomfort, but the workers were not asked to be involved beyond doing what they were told. Occasionally, where union workers were involved, there was violent resistance to such changes. Although this system had occasional problems, it really worked for many decades and formed the basis of traditional management.

So What has Changed?

This book argues that because the rate of technology introduction has and continues to accelerate these traditional management approaches have failed. The competitive world requires that organizations of all types change the technology being used ever more rapidly. Top management does not have the time or knowledge to introduce technological changes at the required pace. This reality has occurred without being well understood or studied. As a result a variety of solutions have involved and there is no "school solution."

Why are Organizations Changing?

Organizations are changing because there is no other option. The ascendance of people at all levels in deciding upon and helping to implement technology change is becoming the norm. These changes are occurring because management has no option – they have to involve employees throughout the organization if the enterprise is going to successfully implement technology changes at the rate required to stay competitive.

MANAGING TECHNOLOGY
DEPENDENT OPERATIONS

What are these New Organizations?

The new organizational paradigm goes by many names – High Performance Organizations, Six Sigma, Baldrige, ISO, etc, etc. The commonality between all of them in the new approach is "power to the worker" - asking people at even the lowest of level to become involved and to contribute to the betterment of the operations.

Once the arguments have been made that rapid technology change is permanently with us and that organizations are moving toward becoming High Performance Organizations, there are a variety of management approaches that require rethinking and likely significant modifications. With this in mind, the authors have comments upon a variety of aspects involved in managing technology dependent operations. The points being made are typically not totally new approaches, but rather interpretations of the historical approaches for the emerging management world. Some of the key variations are presented here.

> Leadership has always required a combination of talents and skills. The new paradigm requires additional focus on motivating and optimizing the performance of people at all levels within the organization. Leadership skills must also be developed through every level of the organization.
>
> Teams and Teamwork are necessary tools for involving people and getting their imput. Organizing, using, and controlling teams has become a critical management skill. This also means that training in team skills are mandatory for the entire workforce.
>
> Projects and Project Management , which were long the purview of specialists, are now required tools at all levels and are used for a wide variety of purposes. Management also has to control a wide

variety of projects that are underway within a given organization and support teams and long term business goals.

Creativity is also found and should be encouraged throughout the organization. Within the new paradigm, the challenge is to encourage creativity at all levels. This has to be done while simultaneously continuing to run "lean and mean" to increase the organizations competitive advantage.

Communications, both verbal and oral, are more important then ever. This challenge is accentuated by the need for communications both up and down as well as to develop clear communications between people with different educational and experiential backgrounds.

Recruiting and Retraining of people has become ever more important. When an organization depends upon its entire workforce, the selection and training of those people is critical. Skills are important, but fitting into the culture is at least as important.

Diversity, of all types, is the norm. To be successful an organization must be able to optimize the contribution of all employees regardless of where they live and their educational background, as well as their race, sex and ethnic backgrounds.

Strategic Planning has typically been the purview of the management team with the balance of the organization simply following through. As responsibility and decision making movies downward within an organization, everyone has to understand the goals and objective of the organization. It is also important that imputes from all levels be included in the planning process.

MANAGING TECHNOLOGY DEPENDENT OPERATIONS

Organizations are not all alike and there is not a single model that works. A small company is not a little version of a large company. Entrepreneurial organizations may be exactly what is needed within a large organization to encourage creativity. Interestingly, the large corporation of yesteryear seems to be the one organizational structure least able to deal with rapid and continuing change.

This book has attempted to provide some tools to anyone who is managing or who wants to manage operations that are technology dependent, and the argument has been made that this includes every organization. There are no correct answers – the solutions will vary between companies and between the people involved.

As a secondary objective, this book has also attempted to suggest some areas where there is a need for additional understanding. Business case studies based on the old management paradigm may no longer be appropriate. It is the editors' strong view that there is need for additional research by business school faculties to better understand and document this new reality.

Simplification and directness is likely to be essential as technical progress gets more and more complex. Technology could become out of control. We cannot let technology automatically generate and spread new technology without control, without human intervention, without management. By reflecting on a technical term, entropy – a measure of disorder that is an ever-increasing property of nature - we can imagine how complex things can get. To deal with it, we must keep our learning process continuous, as simple as possible, and manage to share its load by relying on each other's knowledge and wisdom.

EPILOGUE

So what does the future hold? The answers are not clear. They are still clouded by the "fog of change." There will certainly be additional studies and a variety of publications relating to some of the issues raised in this book. Hopefully, as the future unfolds, there will be clarification on some of the issues raised in this book. At the same time, the one prediction that is certain to be true is that the **future will continue to change.**

Managers who do the best will be those who understand that this is happening and learn to win and succeed in an unstable, ever changing world. Good luck to all of the readers as they start on this quest.

The Editors

Editors/authors:

Donatas Tijunelis is a Professor at the Lake Forest Graduate School of Management MBA program. He teaches as adjunct faculty at Indiana University, National-Louis University, Illinois Institute of Technology and several other universities. He is a member of the International Executive Service Corps and Emeritus Member of Industrial Research Institute. His industrial experience is as Vice President of Viskase Corporation (formerly div. of Union Carbide Corporation). Previously, for many years at the Borg Warner Corporation, he headed manufacturing process research and was the Corporate Liaison for Advance Developments. Don has a number of publications on business and technology management topics. He has a Doctorate in Business Administration and an MS and BS degrees in Chemical Engineering.

Keith E. McKee for many years has been, and is, the Director and President of the Manufacturing Productivity Center as well as the Director of the Manufacturing Programs at the Illinois Institute of Technology. Keith teaches courses in Manufacturing Processes, Quality Control and Resource Management. In the past Keith has held the posts of a Director of Mechanical Design, Product Assurance and Manufacturing Engineering at Andrew Corporation. He was an engineer for the USMC. Dr. McKee published the Manufacturing Productivity Frontiers. He consults to the Chicago Manufacturing Center and Chicago City Colleges. Keith has a PhD, MS, and BS degrees in Civil Engineering from IIT.

Contributing authors:

John Fildes is the CEO of the Packer Group, Inc., a large engineering consulting firm, as well as the CEO of Packer Technologies International, Inc. which focuses on process innovation. John has a PhD in Physical Chemistry form Virginia Polytechnic Institute, BS in Chemistry from Georgetown University.

MANAGING TECHNOLOGY
DEPENDENT OPERATIONS

Richard Hammond is the Vice President of Operations for Harris Corporation, Broadcast Communications Division. Rich has a BS degree in Industrial Engineering from Kansas State University and an MBA from the University of Chicago. Mr. Hammond was Brigadier General in the Army Corps of Engineers where he conducted engineering construction and service programs worldwide.

James P. Nelson is the Director of Materials Research for Illinois Tool Works Corporation (ITW). In the past, for many years, he was Development Manager for AMOCO. Jim has a PhD. and BS in Chemistry from University of Minnesota and University of Illinois, respectively. He is active member of the Technology Management Association of Chicago.

Matthew Puz is the Vice President of Sales & Marketing for Dudek & Bock, a leading metalforming manufacturer. He holds a Bachelor of Science degree in Metallurgy (Metals Science & Engineering) from Pennsylvania State University and an MBA from the University of Michigan.

Tim Ryan is the Director of Business Development for MPR, Inc., an international recruiting and human resource management & consulting firm. Tim has BS degree from University of Illinois with a specialty in entrepreneurship.

Michal Safar is a consultant associated with the Manufacturing Productivity Center and a faculty member of the Manufacturing Technology and Management Program at Illinois Institute of Technology. She has a BA from Butler University and an MA in Library and Information Science from Dominican University.

Eric Spanitz is the President of Synergest, Inc. – a project management consultant and trainer with over sixteen years of experience. He has a BA degree from Northwestern University and is an active member of American Society for Training and Development (ASTD), American Society for Quality (ASQ), and Project Management Institute (PMI).

Index
of
Tables and Figures

Ch 1

Table 1: Evaluating Management Style, p.42

Figure 1: The Project, p.20

Figure 2: Hierarchic Organizations, p.25

Figure 3: Organizational Fit, p.31

Figure 4: Idealized HPO Organization, p.32

Ch 2

Figure 1: Matrix Organization, p.62

Ch 3

Table 1: Operations Strategy Audit, p.75

Table 2: Smart Operations Project Audit, p.76

Table 3: Process & Job Design Audit, p.78

Table 4: Optimizing Process & Product (Service), p.80

Table 5: Review of Total Quality Management, p.81

Table 6: Supply Chain Audit, p.83

Table 7: Forecasting Tools, p.85

Table 8: Resource Planning Review, p.87

Table 9: Inventory Issues, p.88

Table 10: What is going on? , p.91

Table 11: Bench-marking Against The Best, p.92

Figure 1: Components to Build a Process Flowchart, p.77

MANAGING TECHNOLOGY
DEPENDENT OPERATIONS

Ch 4

Table 1: Staff's Circumstances and Leader's Options, p.101

Table 2: Situational Leadership Issues, p.103

Table 3: Origins of Leader's Time Allocation Mis-Allignment, p.110

Table 4: Giving People a Chance to Rise Up - Personal Introspection, p.115

Table 5: Needs-Based Motivation, p.117

Table 6: Motivating Expectations, p.118

Table 7: Survey for the Effectiveness of Teams in Operations, p.122

Table 8: Team Status or Needs, p.123

Table 9: Stages of Progressive Team Development, p.124

Table 10: Questions For Team Membership At The Start, p.125

Table 11: Power Factors Within A Team, p.127

Table 12: Technology "Groupthink" During "Norming", p.129

Table 13: High Performance Teams Checklist, p.138

Figure1: The "Player" to "Coach" Transition, p.113

Ch 5

Figure 1: Components of Innovation in Product Design, p.159

Figure 2: An Innovation Process for Product Design, p.167

Figure 3: Design Project Milestone Chart, p.169

Figure 4: Integrated Product and Process Design for a Metal Casting, p.170

Figure 5: Performance Analysis, p.171

Figure 6: Functional Interaction in the Design and Engineering Phases, p.173

Figure 7: Design Collaboration Web Site, p.177

Figure 8: On-Line Document Library, p.178

Figure 9: On-Line Task Schedule, p.179

Figure 10: Forms-Based Site Management, p.180

Figure 11: RTM I Stiffener Tool Designed With Innovative Process, p.182

Figure 12: Integrated RTM Process Design for I Stiffener, p.183

Figure 13: Innovative LOM RTM Prototype Mold, p.185

Ch 6

Figure 1: The Schedule, p.206

Ch 7

Figure 1: Selection Management Systems, p.238

Figure 2: MPR Competency Mode, p.240

Ch 8

Table 1: Individual Differences on a Basic Level, p.254

Table 2. Rank for Personality Fit With Technical Challenges, p.255

Table 3: Job Content Affecting Attitude, p.258

Table 4: Example of Possible Ethical Decision Alternatives, p.261

Table 5: Knowledge Management in Operations, p.263

Table 6: The Baldrige Based Criteria and Relative Values, p.264

Table 7: Addressing Diversity Problems Between Operational Functions, p.268

Table 8: Cultural Differences and Values in the Workplace, p.277

Table 9: Features of National Economy for Competitive Operations, p.279

Figure 1: Stages of Change Process, p.265

Figure 2: Product Life-Cycle Features, p.267

Figure 3: An Example of Attention and Workload Cycles by Function, p.270

Ch 9

Table 1: Generic Business Strategies for Competition, p.298

Table 2: Project Questions at Varying Business Conditions, p.303

Figure 1: Dominant Views of a Project, p.287

Figure 2: Strategy Roles by Organization Levels in Industry, p.290

Figure 3: The Funnel of Project Evolution, p.292

Figure 4: Strategic Planning Paradigm, p.263

Figure 5: Sources of Strengths and Weaknesses of Operations Management, p.296

Figure 6: The Corporate Strategy Classification, p.297

Figure 7: Leverage and Attention in Project's Life-Cycle, p.319

MANAGING TECHNOLOGY
DEPENDENT OPERATIONS

Ch 10

Table 1: Before Starting Projects: Do not jump at every potential project - carefully choose and organize, p.346

Table 2: When You Open a Project: Ask the right questions — obtain the right answers, p.346

Table 3: Managing Ongoing Projects: Ways to Keep Things Moving, p.347

Table 4: Red Flags for Problem Projects: To Fix or To Kill, p.347

Table 5: Ending Projects: Completing, Killing, and Shelving Projects, p.348

Table 6: Improving Relations With Business Units / Customers: A Win-Win, p.348

Ch 11

Figure 1: Work Breakdown Structure (WBS), p.357

Figure 2: Gantt Chart, p.358

Figure 3: PERT Chart, p.362

Ch12

Table 1: Examining the New Environment, p.390

Table 2: Supply Chain Management Review, p.396

Table 3: Monitoring Discipline, p.401

INDEX

APICS, The Performance Advantage – p. 393
Artificial intelligence – p. 17
Attitudes - p. 257
Baldrige, award – p. 38, 88, 263-264
Bar code, technologies in support of supply chain – p. 395
Behavior Benchmarks – p. 239-245
Bill of materials accuracy (BOM) - p. 398
Business simulation tools, Expected Monetary Value (EMV) - p. 308
Business-to-business, B2B – p. 83
CAD, Industrial Design and Engineering – p. 170-181
Chase, Aguilano and Jacobs – p. 78
Computer aided design tools – 159
Communication, group meetings – p. 223, 168
Communication process, steps – p. 194-197
Communication, presentations – p. 213
Communication, Structured (proposals, report, E-mail) – p. 200
Collaboration software, web site – p. 174 (177)
Culture – p. 119
Cultural differences - p. 277
Core competency, core rigidity – p. 270
Covey – p. 258
Critical Path for the Project – p. 361-362
Deming – p. 80, 387, 395
Dell - 394
DVD, presentation aids - p. 220
Electronic Communication, E- mail – p. 209
Ending Projects - p. 342
Ethical behavior - p. 259
Enterprise Resource Planning (ERP) – p. 86, 391
Evans and Lindsay, Management and Control of Quality - p. 82
Forecasting – p. 84
Frienze, Indiana Labor and Management Council – p. 35
Gaither – p. 78
Gantt chart – p. 169, 358
Globalization, markets/products - p. 273
Goldratt and Cox, The Goal – p. 89, 316
Gordon Ekwall, Swedish Council for Mgmt & Organizational Behavior – p. 34
Groupthink – p. 128
High Performance Organization (HPO) – p. 15, 29

Hierarchical Organizations – p. 15, 25
Hill, International Management – p. 317
Hill and Jones, Strategic Management – p. 75
Heizer and Render – p. 78
Hersey and Blanchard, Situational Leadership – p. 100
IDEO Company – p. 291
Illinois Tool Works, (ITW) – p. 327
Individual differences – p. 254
ISO – p. 38, 81
Integrated Product and Process Design – p. 169, 170
Interdepartmental differences - 265
International Technology Education Association (ITEA) – p. 19
International Management – p. 271
Internet based collaboration – p. 174
Internet technical resources – p. 192
Internet economy – p. 155
InterPROfessional projects (IPRO) – p. 37
Inventory Management, record – p. 88, 399
Innovation, defined – p. 154, 291
Innovation, components – p. 159
Innovation, process, Process Map for Managing Innovation – p. 158-165
Just-in-time – p. 86
Knowledge management – p. 262
Kroeger and Thuessen, Type Talk at Work – p. 256
Leadership, High-Touch – p. 109
Leadership, in practice – p. 103
Leadership, Leader-Doer – p. 111
Leadership, People-Time – p. 110
Leadership, Servant-Leader – p. 106
Leadership, styles – p. 99
Lean (JIT) manufacturing – p. 400
Lincoln Electric – p. 33
Manufacturing Productivity Center – p. 31
Manufacturing Skills Standards Council (MSSC) – p. 36
Martin and Tate – p. 46
Materials Resource Planning (MRP) – p. 86, 391
Matrix organization – p. 62
MBA programs – p. 72
Microsoft Point Team Services – p. 176, (376-378)
Milestone Chart – p. 169, 369
Milne – p. 361
Mintzberg – p. 317

Motivation, "By-the-Book" – p. 116
Motivation, needs based – p. 117
National Association of Purchasing Management – p. 83
National Institute of Standards and Technology (NIST) - p. 265
National Skills Standards Board (NSSB) – p. 36
Netmeeting – p. 181
Oliver Wight ABC Check List – p. 393, 397
Ongoing Projects, managing – p. 337
Online resources, project management - p. 379
Opening a Project, when – p. 333
Operations, definition – p. 64, 73
Operations Management, definition – p. 72
Operations Scheduling – p. 89
Operations Strategy – p. 73
Organization, definition – p. 58
Organizations, world class – p. 15
Packer Technologies International (PTI) – p. 176
Participative management – p. 15
Perception differences – p. 256
PERT Chart – p. 362
Porter – 299, 300
Presentation aids – p. 219
Prioritization, pair-wise example – p. 324
Prioritization, subjective and group techniques – p. 313
Probable Economic Impact (PEI) – p. 311
Problem Projects, red flags from - p. 339
Product Design and Process Selection – p. 77, 78
Process Analysis and Job Design – p. 77, 258
Process improvement – p. 45
Professional Societies, project management-related – p. 381
Project, definition – p. 20, 285-286, 351
Program, definition – p. 352, 284
Project Close-out, Post Project Review – p. 371
Project, leadership – p. 287
Product life-cycle – p. 267
Project Management – p. 76
Project Management Institute (PMI), PMBOK – p. 350, 351
Project management, Funnel of Project Evolution – p. 292
Project Management Software Applications - p. 373, 376, (379-391)
Project, steady-state – p. 25

Prototyping and Testing – p. 171
Purchasing Manager's Index – p. 83
Quality, analysis – p. 82
Quality improvement – p. 45
Quality Management – p. 80
Quality, Six Sigma – p. 38, 82
Quality of Work-life (QWL) – p. 30
Recruiting, Behavior is Key – p. 235
Recruiting, Competency Model – p. 240
Recruiting, "deal breakers" – p. 245
Recruiting, Glossary of Traits – p. 249
Regulations, (OSHA0, (EPA), (NIOSH) – p. 90
Resource Planning and Controlling – p. 85
Royer - p. 318
SAP – p. 86, 393
Scoring techniques, probabilistic models – p. 309
Software selection process, SAP, PeopleSoft, Oracle – p. 393
Standard Operation Procedure (SOP) – p. 15, 22
Starting a Project, before – p. 329
Strategy development, who does what? – p. 288
Strategic Planning Paradigm – p. 293
Supply chain costs – p. 387
SWOT analysis, for projects – p. 295, 302
Teams, building "By-the-Book" – p. 120
Teams, definition – p. 43, 120-121
Teams, High Performance – p. 138
Teams, Performance: "In Practice" – p. 131
Teams, Sample Team Contract – p. 46
Teams, Stages of Team Development – p. 123
Tech Center, tech service – p. 328
Technology, definition – p. 16
Technical information, data management - p. 190
Technology transfer – p. 276
Time-Tracking – p. 110
TQM – p. 38, 80
Traditional management – p. 24
Training, standards – p. 38
Tucker, National Center on Education and Economy – p. 29
Using Microsoft Project, quick summary – p. 376
Utility manager – p. 41
Values in the workplace – p. 277
Value Stream Mapping – p. 391
Wal-Mart - p. 394
Work Breakdown Structure – 357